MINERAL FORMATION AND STRUCTURE
IN THE ELECTROCHEMICAL INDURATION OF WEAK ROCKS

FORMIROVANIE MINERALOV I STRUKTUR
PRI ELEKTROKHIMICHESKOM ZAKREPLENII NEUSTOICHIVYKH GORNYKH POROD

ФОРМИРОВАНИЕ МИНЕРАЛОВ И СТРУКТУР
ПРИ ЭЛЕКТРОХИМИЧЕСКОМ ЗАКРЕПЛЕНИИ НЕУСТОЙЧИВЫХ ГОРНЫХ ПОРОД

MINERAL FORMATION AND STRUCTURE
IN THE
ELECTROCHEMICAL INDURATION OF WEAK ROCKS

By

N. I. Titkov, V. P. Petrov, and A. Ya. Neretina

Authorized translation from the Russian

Springer Science+Business Media, LLC

1965

The Russian text was published by Nauka Press in Moscow in 1964 for
the Institute of Geology and the Exploitation of Fuels of the Academy of
Sciences of the USSR.

**Формирование минералов и структур
при электрохимическом закреплении
неустойчивых горных пород**

*Николай Иосафович Титков,
Валерий Петрович Петров,
Анна Яковлевна Неретина*

Library of Congress Catalog Card Number 65-20210

ISBN 978-1-4899-4907-3   ISBN 978-1-4899-4905-9 (eBook)
DOI 10.1007/978-1-4899-4905-9

© *1965 Springer Science+Business Media New York*

Originally published by Consultants Bureau Enterprises, Inc. in 1965.

CONTENTS

INTRODUCTION

The use of special steel casing pipes as borehole supports entails a considerable expenditure of capital per borehole, reaching as much as 30% of the overall drilling cost.

In gas- and oilfields where the extracted product has a high sulfur content the casing pipes undergo corrosion and must soon be taken out of use; the resulting complications may often be so serious as to necessitate abandonment of the well. It is therefore very important to develop a well design which provides a reliable, simple, and inexpensive shaft support without the use of metal casing pipes.

Great benefits are to be gained from the induration of well wall strata by transforming their properties and compositions (mainly weak clay rocks which readily swell and take up water, thereby complicating drilling). The dc method of indurating rocks is of particular importance; under this treatment weak rocks undergo a change in character, and are thereby greatly strengthened. With the use of special solutions (electrolytes), direct current produces induration of clay, sandstone, and other weak rocks.

Experimental laboratory investigations at the Institute of Geology and Exploitation of Fuels, together with field tests during drilling of shallow wells in the Tatar ASSR, revealed that secondary minerals are formed during the dc induration of weak clay rocks, owing to the use of various electrodes and additions of electrolytes to the aqueous medium; their presence strengthens the rock and increases its water-resistance.

The reactions and processes brought about in the rock by direct current produce various cementing substances (carbonate, silicate, ferruginous), which bind the clay rock particles firmly together. The clay becomes rocklike under water, i.e., the system clay — cementitious material is indurated. The bonding of clay particles and aggregates into a monolithic system is a very complicated process.

Together with changes in the physiochemical nature of the particle surfaces and in chemical composition, the net effect of a dc field is the formation of secondary minerals (and products of accelerated artificial synthesis) and the appearance of a new complex material (transformed rock).

Direct current is characteristically only necessary for initiating crystallization of the minerals formed. In several cases continuation of secondary mineral crystallization and an increase in strength of the clay rock as a whole are observed after the current has been switched off.

The processes of artificial mineral formation are controllable and can be used for investigating natural mineral formation, for strengthening weak rocks and improving their properties (important during well drilling), and in various engineering and hydrotechnical constructions.

The authors would like to thank their colleague of the Institute of Physical Chemistry of the Academy of Sciences of the USSR, Candidate in Chemical Sciences N. N. Serb-Serbina, for her valuable advice.

ESSENTIAL PHYSICOCHEMICAL FEATURES OF THE ELECTROCHEMICAL INDURATION OF CLAYS AND CLAY ROCKS

The electrochemical method of clay and clay rock induration has been fairly extensively discussed in the literature and successfully employed in several industries.

The effect of an electric field on soil was discovered as early as 1808 by Professor F. F. Reiss.

Direct current was first successfully used in the nineteen-thirties for strengthening soil and for the electro-osmotic drying of very wet rock. In 1934 Casagrande [1] found that when an electric current is passed into clay soil via metal electrodes (aluminum and copper), a marked strengthening is observed and some of the soil becomes unwettable. He showed that the extent of the strengthened areas varies with the current strength, electrode shape, and distance between the electrodes. In 1935 Endell [2] found that after direct current had been passed through aqueous pastes (mixtures of sodium bentonite and quartz sand) the moisture content of the paste at the anode fell by 30-40%. The initial mixture in water decomposed in 5 minutes, but after direct current had been passed specimens of these earths did not swell in water for three months.

In 1936 Erlenbach [3] strengthened loams with a moisture content of 37%; he found that soil is strengthened at both the anode and cathode (but to a greater degree at the former). Soil strengthening is not proportional to the amount of current passed. Aluminum electrodes are eaten away during this process. In the USSR, electrochemical induration was carried out by Rzhanitsyn[4, 5], who used Kudinov, Tsaritsyn and Karabulsk clays. He concluded that with the same current densities and duration the induration effect differs for different clays. He modified Casagrande's method and added various salts in the form of electrolytes to the unstrengthened clays. This brought about a considerable improvement, allowing him to strengthen many different types of soil in shorter periods and with lower energy consumptions. These investigations had a marked effect on the subsequent development of the electrochemical method.

In 1937-1940 Pchelintsev and Tolstopyatov experimented with various types of soil in Moscow University; these included screened quartz sand, Glukhovets kaolin, Bureya bentonite, morainic loam, and chernozem. It was found that passage of a direct electric current is always followed by electro-osmosis, leading to drying of the soil and a rise in its temperature (which is always higher at the anode and may reach 100°C or more).

Many Soviet Institutes have carried out research on the strengthening of clay soils by direct current, and similar investigations are now in progress at the Institute of Geology and Exploitation of Fuels.

According to Endell and Hofmann, the changes caused by direct current in clay rocks are the result of exchange reactions between absorbed cations. The replacement of the Na^+ ion by an H^+ or Al^{3+} ion in the absorbed complex may serve as an example; the H ion is formed by the electrolysis of water, the Al^{3+} ion by the electrolytic dissolution of the aluminum anode. They also showed that corrosion of the aluminum anode is a very important factor in rock strengthening because it leads to deposition of aluminum compounds in the clay and formation of the cementing alumogels.

A theoretical explanation (according to the data of Endell and Hofmann) of the mechanism of clay rock induration by direct current has been accepted by many investigators.

Rel'tov and Novikov [6] investigated the cementation of soils by $Al(OH)_3$ and $Fe(OH)_3$ gels in relation to the electrode material, without excluding the possibility of cementation by silicates of bivalent and trivalent metals.

Tolstopyatov [7] expanded the concept of electrochemical soil induration and showed that two types of ionic reaction can take place: 1) physicochemical absorption and exchange reactions between the ions and the soil; 2) chemical reaction between the ions themselves or between the ions and the electrode material. Hydrolysis and binary exchange reactions lead to the formation of cementing deposits of $Al(OH)_3$, $Fe(OH)_3$, $CaCO_3$, $Al_2Mg_5SiO_2O_{10}$, etc.

It will be shown below that a great many physicochemical and chemical processes take place during electrochemical induration. These include solution, oxidation, reduction, and hydrolysis, and are accompanied by physicochemical processes such as temperature change, electro-osmosis, electrophoresis, pore colmatation, coagulation, change in absorption capacity and exchange reactions, migration of electrolytes, and adsorption. The qualitative composition of the clays and clay rock thus undergoes a marked change.

As a result of these processes, artificial secondary mineral cements are formed in the rock, bound by forces of crystallization to the aggregate of clay particles.

In our experiments, dc electrochemical induration of soil was carried out both with and without changes of electrode polarity. The induration results were different in all cases.

Upon examination of the results we investigated separately the strengthened rocks below each electrode between the electrodes (cathodic, anodic, and middle zones), below electrodes with variable polarity (anode-cathode or cathode-anode), etc.

The qualitative changes in the rocks depend primarily on the experimental technique. Induration of the rock in the cathodic zone is due to accumulation of calcium carbonate, aluminum hydroxides (in the form of gibbsite), and iron mineral compounds. In the middle zone the rock is mainly strengthened by active silica and hydrated alumina in the form of allophane, or by accumulation and reaction of hydrated iron and active silica in the form of hisingerite. In those cases where active silica is absent in the central zone, induration is due to precipitation of iron hydroxide in the form of hydrohematite. The anode zone is indurated where it is the site of accumulation of iron hydroxide (in the form of limonite) or in the presence of aluminum hydroxide (in the form of weakly siliceous allophanoid) or markedly ferruginous hisingerite. The peculiarities of these zones are due to the changes in clay composition and formation of new mineral cements, since the zones are closely related to one another by common processes taking place simultaneously.

In the electrochemical induration of wet soil, the most important process is the crystallization of secondary minerals, producing artificial cements (crystallites). On microscopic examination of the hardened rocks it is clearly seen that new mineral compounds, crystallizing in the interstices of the clay aggregate grains, reinforce the bonds between the latter and thus strengthen the clay.

According to [8, 9], electrochemical induration is based on reactions between the electrode material, the electrolyte, and the soluble part of the rock; the temperature, the redox potential Eh, and the composition of the aqueous medium (pH) are important factors. During the experiments, although secondary minerals appear in the clay rock, the initial minerals are often virtually unaffected. It has been noted that for different electrolyte compositions in the same clay, iron is precipitated either with a negative charge in the anode zone or with a positive charge in the cathode zone.

The composition and pH of the liquid phase affect the composition of clay rock. When the composition of the aqueous medium varies, the induration process can be controlled by appropriate additives. Cementation of clay rock — its induration by secondary minerals — can take place in any medium, even one which is inert to an electric current. But a special induration technique must be selected for each rock, particularly if all the rock zones must be strengthened simultaneously.

Secondary minerals may form not only between the individual clay particles and their aggregates, bonding them into a common monolith, but also between the structural units of the rock with formation of bonding cement. Experiments on Kynov argillite from well mud of the Romashka oilfield showed that the argillite was greatly strengthened after current had been passed. Despite the fact that it was not wetted by water at all, its induration was not the result of a change in the rock but of chemical reaction between the electrode material and the added electrolytes, with formation of cementing substances.

Dolomitized clays (in the base of the Upper Kazan substage) of the Novo-Pis'myan region of the Romashka oilfield (outcrops of which are found near Zai-Karatai) have an induration character similar to Kynov argillite.

The most important rock induration factor is the conversion of one form of clay to another, for example, montmorillonitic (Upper Permian) to ferruginous montmorillonite (Fig. 1). This formation process consists in the ferruginization of rocks and the formation of an aluminum gel; these do not separate out as independent formations but are attached to the clay particles.

Other important factors in the electrochemical induration of clays and clay rocks are electro-osmosis, electrophoresis, and exchange reactions, particularly in the case of liquescent clay soil.

Electrophoresis, i.e., the transfer of particles by a dc current, is due to the fact that as the particles of the dispersed phase absorb ions from the aqueous medium, they acquire a charge (potential); the ions may complete the crystal structure of the nucleus, but a diffusion layer of gegenions is located around them. The thickness of the diffusion layer depends on the concentration of electrolytes in the aqueous medium. The higher the concentration of the electrolytes the thinner is the diffusion layer. An increased concentration of electrolytes, for example with a trivalent cation, may change the charge of the micelles, and electrophoresis will then proceed in the reverse direction. Electrophoresis has a particularly marked effect when a direct current is passed through very wet liquescent clay soil, but where the moisture content is low it is virtually absent. Solntsev and co-workers [10] assume that electrophoresis is also possible where little moisture is

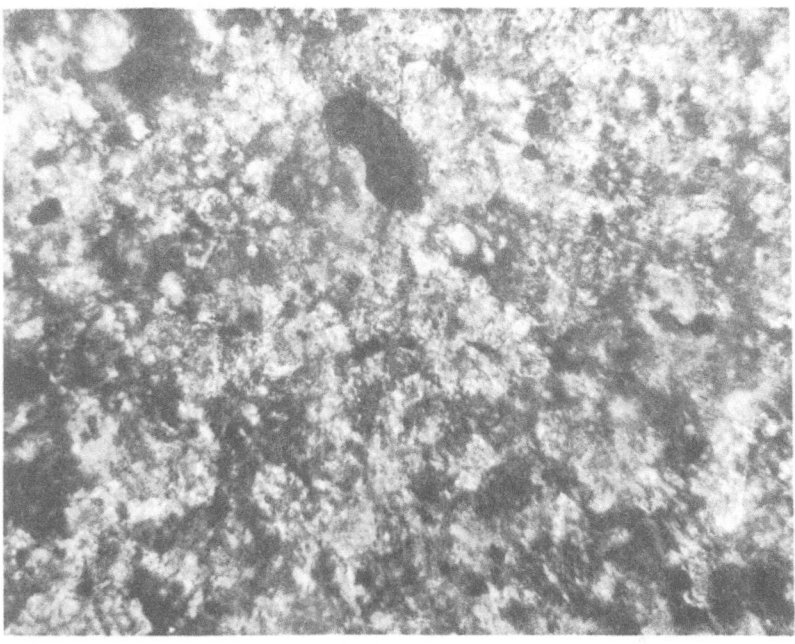

Fig. 1. Rock with secondary ferruginous montmorillonite
(340 ×, crossed nicols).

present, owing to the increase in filtration at the cathode with an increase in the density of the current passed through a carbonate loam with a relatively low moisture content.

If the electrodes are placed in a wet specimen of clay rock at some distance apart, after the current has been switched on, the clay around the anode becomes somewhat drier, while the moisture content near the cathode increases, i.e., water moves from the anode to the cathode. This is observed in all experimental work on electrochemical induration. Under the effect of the dc current, the water in the specimen of clay rock moves from the anode to the cathode and is accumulated there. By the use of a perforated electrode for the cathode, the water may be collected for quantitative and qualitative analysis.

As a result of electro-osmosis the soils dry out and undergo compaction, a phenomenon of some importance in construction operations. Furthermore, electro-osmosis during the electrochemical induration of soils causes the transfer of ions and electrolytes from the anode to the cathode. This may be used for saturating highly dispersed ground with different solutions for induration in cases where the latter cannot be performed by normal methods.

Many investigators attach great importance to exchange reactions occurring in clay rocks when a direct current is passed. This process is normally due to the presence of an Al electrode which is consumed during induration. It is thought that trivalent aluminum replaces monovalent and divalent cations of the absorbent complex of clay rock undergoing induration. Introduction of an aluminum ion into this complex decreases the diffusion layer of the rock's electrochemical potential and coagulates the colloidal clay particles, sometimes leading to a change in sign of the electrokinetic potential at the anode.

Chemical analyses of hardened soil taken from the anode show that aluminum is present not only in the absorbed state but as hydroxide compounds on the surface of the clay particles as well [2, 7]. Furthermore, aluminum hydroxide can serve as a filler. Analysis shows that calcium and magnesium are almost wholly removed from the exchange complex of clay in the anode zone and replaced by hydrogen and aluminum. The displaced calcium is accumulated in the zone adjoining the cathode. It is present both absorbed in the exchange complex and as hydrates. Investigations show that calcification occurs near the cathode and the induration is increased. The strengthening of the soil is due to cementation, resulting from calcium and magnesium hydroxides and the subsequent carbonation of these hydroxides.

During the electrochemical induration of clay rock, elements playing the role of a coagulator may appear in the anode zone for reasons other than dissolution of the electrode substance. A markedly acid reaction in the vicinity of the anode assists the appearance of aluminum and iron in the solution, these being extracted from the ground itself. This was confirmed by Tolstopyatov's experiments on chernozem, strengthened by the use of carbon electrodes.

The effect exerted on induration by introduction of the aluminum ion into the absorption complex, and by accumulation of aluminum hydroxides and basic salts in the anode zone, led some investigators to suggest replacement of aluminum electrodes by iron ones.

In [6] successful experiments are described concerning the use of an iron anode for the electrochemical induration of ground. Iron anodes sometimes give better results than aluminum ones; this is the case when sodium, which facilitates precipitation of limonite in the anode zone, is present in the aqueous medium.

Exchange reactions in rock take place in dilute solutions at low current density; if the conditions are otherwise, the exchange reactions cease. In this event, crystallization processes predominate with formation of new compounds which act as the cement.

CHARACTERISTICS OF THE INITIAL CLAYS

Investigations into the effect of direct current on the chemical and mineralogical composition of clays were carried out on six types of clay, three monomineralic and three polymineralic.

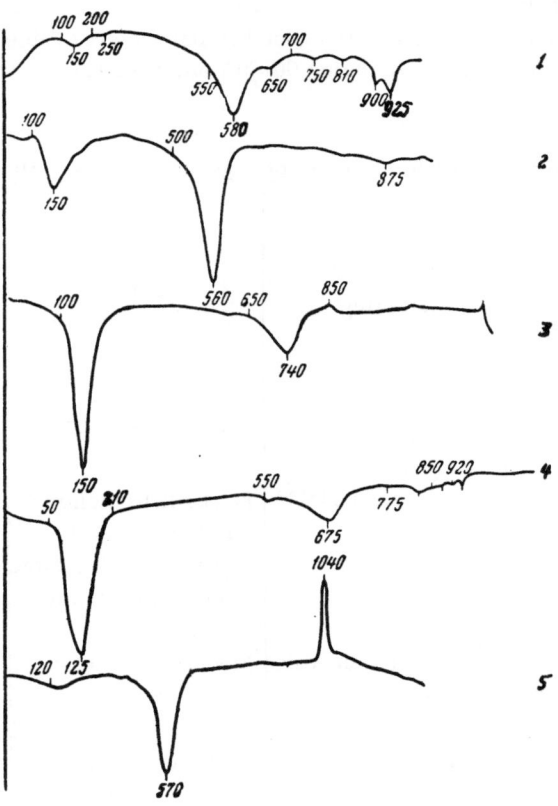

Fig. 2. Differential thermal curves of the initial monomineralic clays. Chasov-Yar monothermite: 1) white; 2) light gray; bentonite (Oglanly): 3) 1st grade; 4) 2nd grade; 5) kaolin.

1. Fire clay (monothermite) from the Chasov-Yar deposit, which is located at the watershed between the Krasnyi Terets and the Stupka.

The fire clay is subordinate to Paleogene deposits which lie nonuniformly on the eroded surface of Mesozoic and Paleozoic rocks. Together with the fire clay, the Paleogene deposits have suffered considerable erosion. The cover thickness of the clay varies from 1 to 40 m. The cover rocks are white and yellow quartz sands of the Poltava stage and brown Quaternary loams.

Chasov-Yar clay has the following composition: losses on baking not less than 7.5-8%; SiO_2 not more than 54-59%; Al_2O_3 not more than 28-32%; Fe_2O_3 not more than 1.5-2.3%.

This clay is white or light gray, plastic, greasy to the touch, and contains yellowish-brown spots. Under the microscope it is fine-grained pelitomorphic, the structure is flaky, and rosette-like segregations are observed. For the monothermite $\gamma = 1.569$, and $\alpha = 1.548$. Individual grains of calcite and hydromuscovite with $\gamma = 1.570$ are observed. Very fine tetragons, evidently sulfate (löwigite) are observed in the admixture; some parts of the monothermite have a quartz admixture.

Thermal analysis of the white area of monothermite showed endoeffects at 150, 580°C and at 900, 925°C (Fig. 2; 1).

Thermal analysis of the light gray area showed endoeffects at 150, 560°C, characteristic of monothermite, and an endoeffect at 875°C (Fig. 2; 2).

The high-temperature endoeffects tend to indicate that the clay contains an admixture of carbonates, dolomite (Fig. 2, curve 1) and calcite (Fig. 2, curve 2). The grain composition of the clay and the character of the water extract are given in Tables 1 and 2.

2. Oglanly clays (true bentonite deposit). The deposit is located in West Turkmenistan and consists of a series of tripoli and marl, with a band of Paleogene bentonite clays (Table 3). The bentonite series is conformable with the tripoli, the angle of dip being 60-70°, and is correlated with the north flank of the large Balkhan anticline.

TABLE 1. Characteristics of Aqueous Extracts of the Clays

Geological age	Site of clay sampling	Clay	pH of the aqueous extract	Absorption capacity, meq/100 g clay	Composition of the exchange cations	
					Ca + Mg	Na + K
Upper Permian	Aleksandrov site, Tuimazy	Montmorillonite	7	26,1	22.3	3.8
Upper Jurassic	Quarry near Podol'sk	Ferruginous montmorillonite (poorly graded, with a large amount of humic acids)	6.4*	52.5	52,1	0.4
Paleogene	Chasov-Yar deposit	Montmorillonite	7.4	18.2	16.7	1.5
Mesozoic erosion crust	Prosyanov deposit	Kaolin (primary preparation product)	8.8†	6.5	6.0	0.5
Upper Permian	Oglanly deposit	Montmorillonite	7.5	60.7	14.5	46.2

*The reduction in the pH of the aqueous extract of ferruginous montmorillonite is evidently due to the presence of a large number of other minerals in the rock.

†The high pH of the aqueous extract of kaolin is due to the preparation process employed, which involves the use of NaOH.

TABLE 2. Grain Composition and Swelling of Clays (%)

Geological age	Clay	Fraction, mm								Swelling, %
		More than 0.5	0.5–0.25	0.25–0.1	0.1–0.05	0.05–0.01	0.01–0.002	0.002–0.001	0.001	
Upper Permian	Montmorillonitic	0.14	0.51	1.31	22.62	2.69	37.71	2.70	32.32	38.8
Upper Jurassic	Ferruginous montmorillonite (enriched with terrigenous minerals)	–	29.06	23.52	4.15	–	28.06	13.68	1.53	35.9
Mesozoic erosion crust	Kaolin (washed)	–	–	1.7	18.9	7.9	6.4	30.2	34.9	16.5
Paleogene	Bentonite	–	–	0.57	2.55	12.54	15.92	11.40	56.92	–
Upper Devonian (Kynov formation)	Hydromica (with an admixture of montmorillonite)	–	–	–	17.35	8.05	27.56	16.12	30.82	14.79

TABLE 3. Chemical Composition of Oglanly Clay

(% in terms of dry substance)

Components	Sample				
	1	2	3	4	5
SiO_2	73.50	75.09	71.12	73.86	72.05
Fe_2O_3	2.31	2.02	1.99	2.77	7.82
Al_2O_3	13.14	9.87	12.15	13.10	7.99
CaO	1.23	0.98	1.60	1.02	1.60
MgO	2.91	2.30	2.51	3.03	2.19
K_2O		0.94		0.73	
Na_2O	3.11	3.88	4.20	1.44	2.42
SO_3	0.99	0.56	0.57	0.87	0.36
Losses on drying	2.53	3.83	5.30	2.73	6.42
Total	99.94	99.47	99.96	99.69	100.87
Hygroscopic moisture	14.67	15.61	10.86	15.00	9.40

Cover rocks in contact with the bentonite series in the south include a pebble bed (conglomerate) with a thickness varying from 5-10 m to 15-18 m, and loess-like loams 1.5-3 m thick.

The overall thickness of the bentonite band varies from 9.5 to 23.7 m.

Microscopic investigations of samples of Oglanly bentonite show it to consist largely of finely dispersed material — primarily montmorillonite, hydromica, and some biotite and calcite.

The refractive index of montmorillonite from primary grade bentonite clay is $\gamma = 1.522$, $\alpha = 1.496$, while for the second grade it is $\gamma = 1.521$, $\alpha = 1.495$.

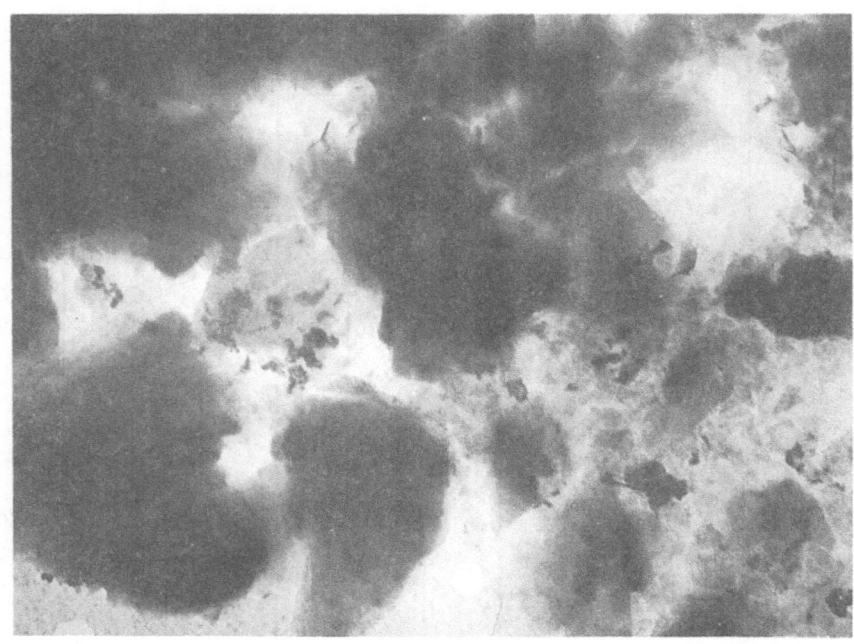

Fig. 3. Electron micrograph of Oglanly bentonite (× 12,500).

Thermal analysis of primary-grade bentonite showed endoeffects at 100-150°C and 740°C (Fig. 2, curve 3); the second grade showed endoeffects at 50-125°C and 675°C. The two small endoeffects at 550°C and 920°C indicate an admixture of hydromica and calcite (Fig. 2, curve 4). The electron micrograph of Oglanly bentonite is shown in Fig. 3.

3. Primary kaolin of the Prosyanaya deposit. This deposit, the largest kaolin deposit in the Ukraine, is located in the Dnepropetrovsk region, where it was formed by the erosion of massive migmatites and granite with a thickness of more than 50 m. The thickness of the overburden (sand and clay) varies from 0.5 to 26 m. The chemical composition of primary-grade washed kaolin (in %) varies within the limits:

SiO_2	TiO_2	Al_2O_3	Fe_2O_3	CaO
46.07—46.85	0.25—0.35	37.38—39.83	0.3—0.73	0.15—0.56

The losses on baking are 13.34-13.97.

The kaolin clay is white, washed, structureless, and greasy to the touch. A microsection cannot be made because the kaolin disintegrates into individual particles.

In immersion under the microscope the kaolin reveals discrete particles or haphazardly aggregated ones; the refractive index is $\gamma = 1.567$, $\alpha = 1.562$. A small admixture of calcite, quartz, and occasional inclusions of plant residue are observed.

Thermal analysis of these specimens revealed the presence of pure kaolin, indicated by the slight endoeffect at 120°C, the marked endoeffect at 570°C, and the exoeffect at 1040°C (Fig. 2, curve 5).

4. Upper Permian (Kazan stage) clay from the Aleksandrov area (Tuimazy) is greenish gray, compact, markedly calciferous, greasy to the touch, and is readily wetted in water.

Under the microscope thin sections of the clay have a carbonate-montmorillonite composition with an admixture of hydromica and a lepidoblastic-pelitomorphic texture; the structure is not bedded.

The principal minerals forming the clay are montmorillonite, pelitomorphic calcite, and hydromica. Small amounts of chlorite, biotite, muscovite, limonite, quartz, and feldspar are observed together with

16

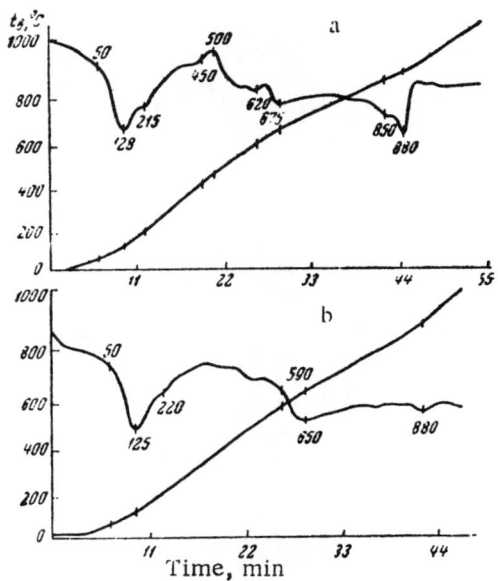

Fig. 4. Differential thermal curves of Upper Permian clay: a) initial; b) treated with 3% HCl.

occasional grains of chalcedony, kaolinite, hematite, zircon, and titanite. The refractive index of the montmorillonite groundmass is 1.540; another montmorillonitic modification is present which is colorless and which has refractive index γ = 1.528 and α = 1.509 — this was evidently formed from the hydromuscovite. The latter has refractive index γ = 1.559 and α = 1.543, while the refractive index of muscovite is γ = 1.588 and α = 1.553.

Thermal analysis of these specimens (Fig. 4a) showed that the clay has a montmorillonitic composition with absorbed Ca and Mg cations, indicated by the endoeffects at 50-128, 215, and 675°C, while the endoeffect at 880°C shows the presence of calcite; this can be removed by treating the specimen with 3% hydrochloric acid in the cold (Fig. 4b).

The overall chemical composition of the clay (in %) is given below:

Components	Upper Permian Montmorillonite clay	Jurassic Ferruginous montmorillonite	Components	Upper Permian Montmorillonite clay	Jurassic Ferruginous montmorillonite
SiO_2	50.66	55.55	Na_2O	0.88	0.35
Al_2O_3	14.38	21.08	K_2O	2.48	2.59
Fe_2O_3	6.68	5.95	CO_2	6.77	Traces
CaO	9.35	1.54	H_2O	3.80	7.60
MgO	4.65	1.81			

A micrograph of the initial Upper Permian clay is shown in Fig. 5.

5. Upper Jurassic (Oxfordian stage) clay from the Podol'sk quarry. This clay is dark brown, friable, lumpy, weakly cemented in places, markedly micaceous, and readily wetted in water.

Microscopic examination of this clay reveals a hydromicaceous-montmorillonitic composition with a high content of muscovite and siltstone particles. The structure is lepidoblastic and the texture inequigranular, the distinctive feature being the presence of areas with different siltstone particle contents.

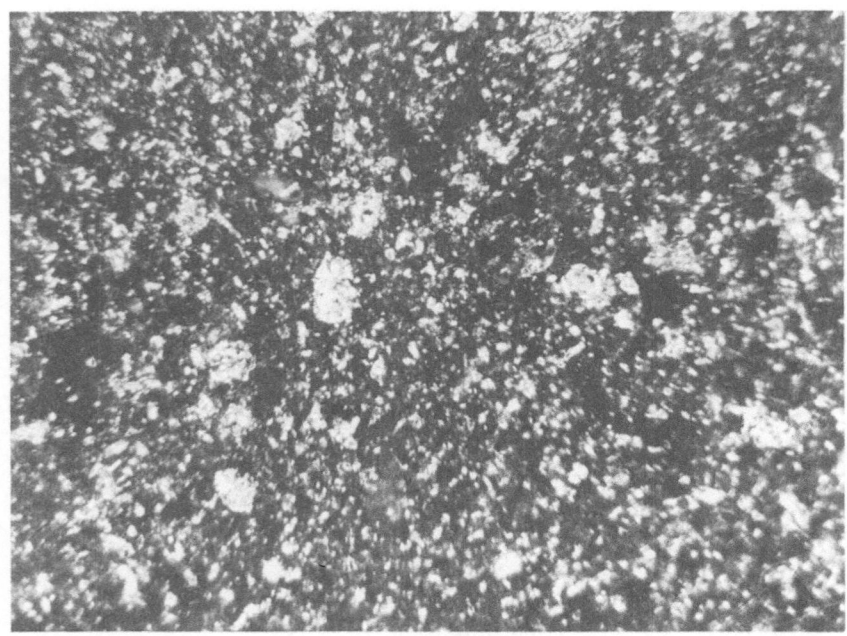

Fig. 5. Initial Upper Permian clay (× 340, crossed nicols).

The principal minerals of which the clay is composed are ferruginous montmorillonite and hydromica. Large amounts of muscovite, quartz, feldspar, jarosite, and plant residue were present in the specimens investigated. Grains of chalcedony, kaolinite, allophane, biotite, calcite, pyrite, magnetite, gypsum, epidote, zoisite, zircon, and hypersthene are sometimes observed. The chalcedony was observed as rolled pebbles, and the kaolinite and allophane as irregular fragments.

The characteristic feature of this Jurassic clay is the presence of up to 6.2% humic acids.

The refractive index of the brown montmorillonite of this clay is greater than 1.540, and that of colorless montmorillonite is 1.516.

Thermal analysis of these specimens showed that the composition of the clay is ferruginous montmorillonite, which is indicated by the endoeffects at 75-140 and 550-600°C and the exoeffect at 865°C (Fig. 6).

Our investigations are in substantial agreement with the data of Nikolaevskii [11] and Feodot'ev et al. [12].

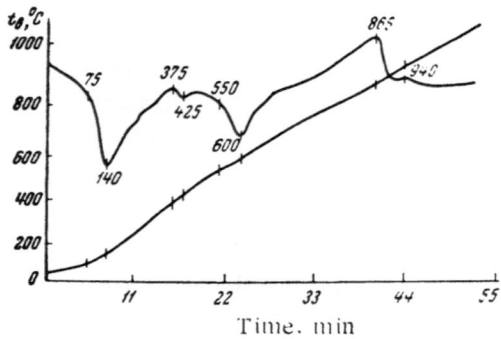

Fig. 6. Differential thermal curve of Upper Jurassic clay.

6. Kynov clay from the Upper Devonian argillite series of the Romashka oilfield (Tatar ASSR).

The clay varies in color from dark to light green in some areas and from dark brown to reddish brown (and is noncalciferous) in others; light-brown sideritic lenses and spots are everywhere observed.

The clay is fine-grained and lepidoblastic, consisting mainly of hydromica with an admixture of montmorillonite, fine-grained siderite, and iron hydroxides.

Thermal investigations of 13 specimens of Kynov clay confirmed the mineralogical composition shown under the microscope; this is hydromica clay with a considerable admixture of montmorillonite in some areas; nearly all speci-

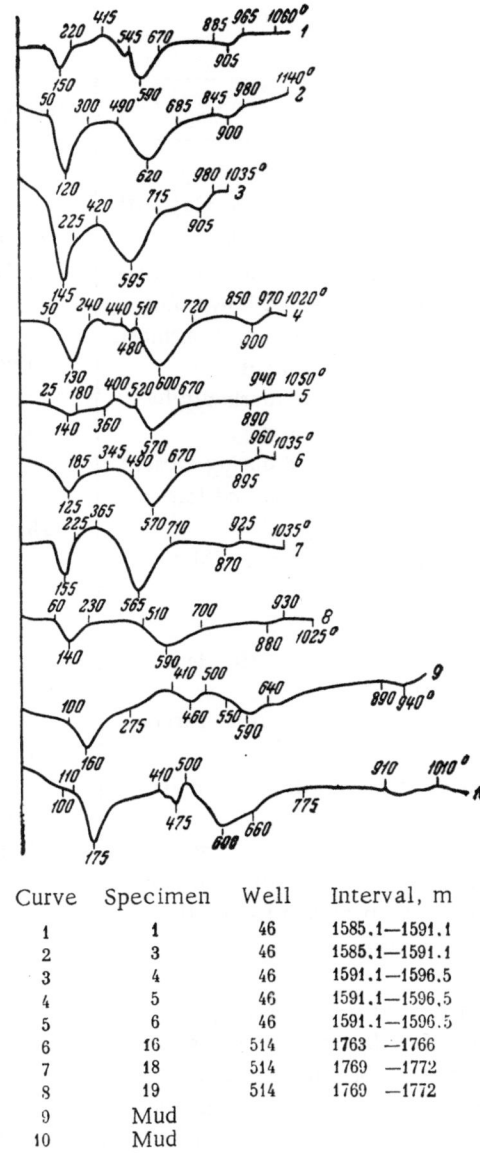

Curve	Specimen	Well	Interval, m
1	1	46	1585.1—1591.1
2	3	46	1585.1—1591.1
3	4	46	1591.1—1596.5
4	5	46	1591.1—1596.5
5	6	46	1591.1—1596.5
6	16	514	1763 —1766
7	18	514	1769 —1772
8	19	514	1769 —1772
9	Mud		
10	Mud		

Fig. 7. Differential thermal curves of Kynov clay.

mens contain iron hydroxides and thermal observations confirmed the presence of siderite in some.

It was found that argillite of different horizons differs in its mineralogy, a fact which is clearly indicated on the heating curves. Figure 7 gives the original curves of argillite from different levels (see curves 1-8); a general similarity between the argillites is observed in these curves.

For this investigation it was therefore desirable to employ an average sample; in our opinion, drilling mud was a suitable choice (see Fig. 7, curves 9-10).

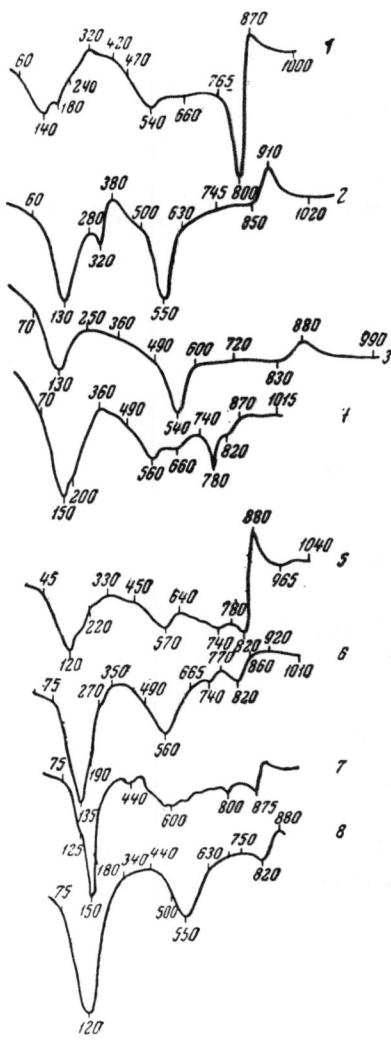

Fig. 8. Differential thermal curves of the initial clays used in the experiments. Ferruginous montmorillonite: 1) Bogdanovsk; 2) Baiguzino; 3) hydromica—montmorillonitic (Baiguzino); 4) sandy montmorillonitic (Tel'men-Elga); 5) Sultangul ferruginous montmorillonite; 6) montmorillonitic clay with an admixture of "Krasnoyarka" hydromica; 7) Kiryushin montmorillonitic clay with gypsum; 8) ferruginous montmorillonite from the Kokhan site.

TABLE 4. Principal Characteristics of Clay Rocks

Region and deposit	Geological age	Rock	Swelling, %
Ishimbay			
Bogdanovka	Quaternary deposits	Carbonate—montmorillonitic clay with about 10% siltstone particles	18.15
Kazankovka	Tertiary deposits	Montmorillonitic clay with hydromica	75.42
		Montmorillonitic clay with hydromica	93.6
Baiguzino		Clay formed during erosion, montmorillonitic, markedly cemented	34.28
		Montmorillonitic clay, markedly but nonuniformly cemented	30.59
Termen'-Elga	Upper Permian, Ufa deposits	Loam (clay part formed during erosion, consisting of hydromica and ferruginized clay)	25.16
Smakaevo		Reddish-brown ferruginized dolomitized clay	
Buguruslan region			
Sultangulovo	Quaternary deposits	Carbonate—montmorillonitic clay, well washed	46.94
Krasnoyarsk		Montmorillonitic, hydromica, carbonate clay	75.6
Kiryushin		Carbonate — Montmorillonitic, markedly cemented	45.64
Krasnoyarka	Quaternary deposits	Montmorillonitic clay with hydromica	62.89
Mukhanov (from the Kakhan site		Montmorillonitic clay with an admixture of hydromica and finely dispersed gypsum	118.92
Frunze region			
Moscow Metro	Carboniferous	Carbonate clay; the greenish-gray areas consist of carbonate—montmorillonite clay; pinkish-brown areas of carbonate—hydromica with kaolinite and a small admixture of montmorillonite	19.44
Moscow Metro	Jurassic	Dark gray, nearly black siltstone	12.7
		Sand with sandstone fragments	12.89
		Dark gray sandstone; the argillaceous part consists of hydromica and ferruginous montmorillonite	18.71

In addition to the main investigations on the six clay types, successful induration experiments were also performed on the clays whose characteristics are given in Table 4 and Fig. 8.

The large number of specimens of clay and clay rocks of different composition employed in the laboratory investigations was necessary to develop the most effective method of electrochemical induration of these rocks.

EXPERIMENTAL PROCEDURE EMPLOYED FOR
THE ELECTROCHEMICAL INDURATION OF CLAYS

Before induration the clays or clay rocks were wetted and then rammed into a Plexiglas bath. The water ratio of the prepared clay depended on the composition and quality of the latter.

In these experiments various amounts of the following were added: $NaCl$, $CaCl_2$, $MgCl_2$, $AlCl_3$, $FeCl_3$, $Na_2O \cdot SiO_2$, stratal water, $CaCO_3$, $NaOH$, a saturated solution of $Ca(SO_4)_3 \cdot 18H_2O$, a 1% suspension of $Al(OH)_3$, and carboxymethylcellulose. These last four were added only in a few cases; those chiefly employed were stratal water, water glass, and calcium chloride (which give the best induration results). Calcium carbonate was added as the dry powder (15% wt. in terms of montmorillonitic clay), as it yields good induration results in the cathode zone. Sodium hydroxide was used in investigations on the movement and precipitation of iron in the cathode and anode zones.

Aluminum, iron, and graphite electrodes were used. For the most part we employed laminated electrodes placed on the surface of the rammed and moistened rock, the interelectrode distance being 3-4 cm. The country rock was represented by Plexiglas baths.

To obtain a uniformly distributed electric field, the electrodes were placed in the same plane in the same numbers on both sides of the soil or rock to be indurated.

A direct electric current was fed to the electrodes via a rectifier at up to 150 V for soils from 25-80 V. Soils are, of course, conductors and electrolytes with respect to an electric current. In moist soils current transfer is due to movement of particles and ions to both electrodes; the current density is a decisive factor in the induration of rock.

If the current density does not exceed 1 ma/cm^2 during the experiments, only coagulation will occur; when the current is switched off the process will continue by colloid aging and syneresis; in this case the minerals will remain in the form of powders, having no bonds with the rock to be strengthened. Crystallization processes also take place at a current density > 5 ma/cm^2, where the temperature reached during electroinduration is also important.

Electrochemical induration by direct current is carried out by two methods, i.e., with and without a change in polarity. Where there is no change in polarity, two distinct zones are formed — a "cathode" and an "anode" zone. Where the polarity is changed, zonality is largely eliminated, the zones being arbitrarily described as cathode—anode and anode—cathode. During the laboratory investigations the current acted on the rock for a period varying from 5 to 100 hours. After completing the electrochemical induration of experimental specimens of clays and clay rocks, we investigated individual layers of the strengthened rock.

The first surface layer of the rock was 0-20 mm, the second layer 20-40 mm, and the third 40-60 mm. During induration the temperatures of the rock and liquid phase reached 80 and 120°C, respectively. The pH changed sharply from 1 to 4 (the pH is 1 at the completion of induration) in the anode zone, and from 7 to 14 (10-12 at the end of the process) in the cathode zone.

After completion of an experiment, the pH in the anode and cathode zones were measured on a LP-5 potentiometer with paired antimony and calomel electrodes; the pH of the aqueous extracts was determined by means of a glass electrode. After these measurements the specimen was removed from the baths and its surface was then divided into two actual parts. One part was assumed to be the cathode zone, and the other the anode zone; samples of these zones were taken for strength tests.

During the electrochemical induration of clay rocks, coagulation and crystallization structures are formed. With complete saturation by water (for 7 days) specimens with a coagulation structure decompose. The crystallization structure is determined in the experimental installation (a four-ton hydraulic press) after the specimen has been completely saturated.

With coagulation structure the rock is only compacted, reversible processes being a characteristic feature.

In this case, when the specimens are kept in the air-dry state, colloid aging and syneresis occur. Hydroxides are formed which disintegrate in water.

Crystallization strengthens the rock, and the processes are irreversible: the rock does not become sodden when completely saturated with water. The most important crystallization process occurring during electrochemical induration is the formation of artificial minerals firmly attached to the whole rock mass.

The induration of clay rock as a result of secondary mineral formation is due to the fact that the nucleating centers of the crystals formed during electrochemical treatment continue to grow actively from the walls of the pore cavities and fissures, thereby ensuring adhesion of the rock.

The formation of artificial gibbsite and allophane is indicated in [12]. However, formation of these minerals takes place without a bond between the individual crystals, whereas the artificial minerals formed during the electrochemical induration of clay rocks are cements; as regards its strength properties, the mono-mineral formed on the surface corresponds to the natural mineral.

PROCEDURE FOR INVESTIGATING ELECTROCHEMICALLY INDURATED CLAYS

The following procedure was employed for investigating the mineral and petrographic properties of electrochemically indurated clay:

(1) In polished sections investigations were made of the general petrographic composition, texture, and structure of the clays, and a preliminary identification of the minerals was undertaken.

(2) Using the immersion method, a closer study was made of the nature of macromineral grains.

(3) Thermal analyses were performed, allowing an accurate assessment of the nature of the bulk of the rock.

(4) The nature of the dispersed minerals was established by an X-ray investigation.

(5) Chemical analyses, supplemented by other methods, were performed.

One of the main purposes of the investigations was to make a comparison of data on the mineral-petrographic composition and the physicomechanical properties of the specimens.

The final assessment of the mineral composition of the clays, and of the changes undergone, was made after comparing all the data obtained by the investigations as a whole. Where any doubt was felt, the investigations were repeated until the results agreed.

Mechanical Analysis of the Clays

The main task was the differentiation of the clay material into groups (fractions) of different grain size, so as to determine the structural features and find the difference in the compositions of the individual fractions.

Preparation of the Specimen for Analysis

This consisted in preliminary mechanical disintegration of the rock, with destruction of the bonding cement in some cases. Additives (NH_4OH, $Na_4P_2O_7$, $NaCO_3$, etc.) were sometimes employed to assist disintegration.

To remove calcite from the rock, we treated it during disintegration with 3% HCl in the cold, then transferred it to a fine filter, washed it with distilled water to remove chloride ions, and air-dried it in the pump. If the dried rock shrunk, further disintegration was effected in which the clay was completely mixed. The average sample obtained in this way was used for immersion investigations, determination of the amount of minerals, chemical analysis, and X-ray investigations.

The compact rocks were prepared in a special way for the grain size analysis. Maximum separation of the particles was obtained by employing a sodium pyrophosphate solution as separating additive. By adding 1 ml of 4% solution per gram of rock analyzed we obtained a stable noncoagulating suspension which could be transferred via a pipet.

The original clays were investigated under the microscope before treatment with hydrochloric acid and bases, and the composition of their aqueous extracts was determined.

We prepared the analytical specimens of strengthened clay in a strict sequence. First the specimens were studied microscopically, and very distinct areas with accumulations of secondary minerals were distinguished. We also studied the textural and structural characteristics.

Where the cement in the strengthened rock was calcite, the rock was treated by the normal method for the initial rocks. But if the strengthened rock contained a secondary mineral with some other composition, insoluble in weak acids and bases (for example, gibbsite, limonite, hydrogoethite, lepidocrocite, and hydrohematite), the rock was crushed. The fragments were first studied under a binocular lens so as to select the individual minerals, which were then identified under a microscope by the immersion method.

Thin sections were made from the strengthened rock lying perpendicular to the plane of action of the current, using anhydrous glycerine to prevent the loss of newly formed water-soluble minerals.

Investigation of Thin Sections under the Microscope

Here we studied the general petrographic composition, texture, and structure of the original change and made a preliminary identification of the minerals.

Study of these thin sections allowed us to determine the general composition of the strengthened rock, the formation of artificial textures and structures, and the location and composition of the artificial cements.

Immersion Investigation Procedure

The immersion method of optical investigation of natural and artificial minerals, which allows one to determine the refractive indices of the different minerals under the microscope, is widely used in the Geological Survey and in industry.

It is important for strengthened rocks because the clay particles are sometimes so small that, of all the optical constants, only the refractive index can be determined.

The refractive indices are mainly determined from observations of the Becke line, but we also made observations on the color effect, indirect illumination, and the shagreen surface of the grains.

The color reaction method for determining the refractive index is based on the fact that when the indices of the liquid and the mineral are close, but their degrees of dispersion different, the grains appear colored. To obtain the reaction we used a lateral parallel beam obtained by partial diaphragming. Lateral illumination conditions were established by darkening the visual field near the grain with an opaque plate placed in the slot channel of the microscope tube. Where the mineral's refractive index was greater than that of the liquid, a dark band appeared on the grain side facing the darkened area, and vice versa. Where the refractive indices were equal (very often not exactly equal), the mineral turned red and blue at the ends.

From a careful examination of grains immersed in a liquid with a similar refractive index, we could readily discern a pinkish dispersion color for minerals with a lower index than the liquid, and a bluish tint when the index was higher.

For fine grains the Becke effect is observed not in the form of a line, but rather as illumination and darkening of the whole grain. Where the mineral had a higher index than the liquid, when the tube was raised the grain was completely illuminated and its outlines seemed to "constrict." Where the mineral had a lower index the grain was darkened and its outlines enlarged. In these cases we could measure particles about 0.001 mm in size.

The normal procedure for making immersion preparations was slightly modified for the strengthened clays. The powdered rock was first ground in the immersion liquid in a mortar and distributed uniformly over the whole preparation by pressing repeatedly on the cover glass.

In immersion preparations measurement of the main refractive indices can be accompanied by determination of the mineral's color, birefringence, extinction, optical sign, pleochroism, etc; these constants are often sufficient to identify the mineral.

Thermal Investigations

The initial rock was generally investigated in the form of an average sample without prior treatment; however, rock with a high content of carbonates, calcite, and fine-grained dolomite was treated with cold 3% HCl and dried in air.

After thermal investigation, the untreated strengthened rock was washed with hot distilled water and dried; the thermal curve was then redetermined. This process allowed us both to record the effects of water-soluble salts in the rock and to study the composition of the dry residue of the aqueous extract from the same clay.

For the thermal investigations we used a photorecording pyrometer of the N. S. Kurnakov system, with roasted alumina as the standard.

Chemical Analysis

Bulk chemical analyses of the strengthened rocks were performed in both the anode and cathode zones. We also analyzed specimens of secondary minerals. These were prepared for analysis in several ways; in some cases the specimens were dressed by hand under a magnifying glass, in others by centrifuging in heavy media. Minerals with low densities were dressed with a mixture of bromoform and benzene.

Chemical analyses for the determination of the molecular formulas of the minerals were performed on monomineralic samples.

The bulk chemical analyses were analyzed in terms of the percentage ratios of individual oxides in the strengthened rocks, compared with the original ones.

X-Ray Analysis

In many cases X-ray structural investigation provides a more accurate diagnosis of the minerals, facilitates assessment of the grain size of the secondary minerals, and enables one to detect the growth of their crystals (which sometimes continues after the current has been stopped).

The specimens for this analysis were prepared in two ways: the monomineral samples were selected either by hand with the aid of a magnifying glass, or by centrifuging.

The X-ray photographs were taken by the Debye method in a 57.3-mm diameter chamber, using Cu $\alpha_{1,2}$ and Fe $\alpha_{1,2}$ radiation; the specimen diameter was 0.5 mm.

After completion of the measurements (making a correction for specimen absorption) the powder patterns were interpreted from tables in the usual manner.

DESCRIPTION OF THE SECONDARY MINERALS

Let us examine the formation and characteristics of the most important secondary minerals.

Gibbsite (hydrargillite), $Al_2O_3 \cdot 3H_2O$, is formed in all clays by treatment with direct current, using aluminum electrodes in a strongly alkaline medium (pH 10-12). The presence of calcite together with gibbsite is a characteristic feature (according to [13], this is also characteristic of natural gibbsite).

If the current acts on clay rock for 30 hours, finely dispersed alumina monohydrate with a crystal size of about 0.001 mm is precipitated. With longer treatment, distinct crystals of gibbsite (0.01 mm or more) are deposited in the form of an overgrowth cement. Maximum accumulation of gibbsite is observed in carbonate—montmorillonite clay.

Gibbsite forms a cement in the strengthened rock and dense white crusts on its surface, their thickness depending on the duration of the current.

The microscopic picture of gibbsite is very typical: a white color (sometimes with a bluish tint), oblique extinction, and in many cases twinned crystals. The refractive index is $\gamma = 1.588-1.591$ and $\alpha = 1.571-1.576$ (the first values relating to the finely dispersed phase).

The chemical analysis of secondary gibbsite is as follows (%): $SiO_2 - 0.32$; $Al_2O_3 - 68.82$; $Fe_2O_3 - 0.51$; CaO -0.68; $Mg - 0.12$; $H_2O^- - 2.79$; $H_2O^+ - 27.11$; total 100.35. It is thus observed to be very pure. But molecular recalculation reveals a certain deficiency of water ($Al_2O_3 \cdot 2.47 H_2O$) in the formula for secondary gibbsite, which may be due to the presence of the monohydrate $Al_2O_3 \cdot H_2O$. The term "gibbsite monohydrate" is an arbitrary one; the substance is probably actually a mixture of gibbsite trihydrate and electrode aluminum (further study is required to clear up this point).

X-ray structural analysis of artificial gibbsite (Table 5) shows that the interplanar spacings of this material coincide with those of natural [14] and artificial [15] gibbsite, within the limits of error.

By crystallizing in the interstices between the grains of the clay aggregates, gibbsite strengthens the clays and gives a distinctive structure to current-treated rock.

From our experiments, the type of heating curve obtained for gibbsite with a small admixture of montmorillonitic clay has endoeffects at 250-275-300°C and 470-510°C (Fig. 9a).

An incompletely dressed specimen of artificial gibbsite showed endoeffects at 210-240-250-300°C and 425-500°C (Fig. 9b). In artificial gibbsite the third endoeffect is displaced somewhat toward lower temperatures (470-510°C instead of 490-550°C). The artificial gibbsite obtained by electrochemical induration has a distinct third endoeffect.

TABLE 5. Comparative Data on the Interplanar Spacings of Different Gibbsites

Artificial gibbsite formed during induration		Artificial gibbsite, according to Kitaigorodskii [14]		Native gibbsite, according to Rooksby [15]		Artificial gibbsite formed during induration		Native gibbsite, according to Kitaigorodskii [14]		Artificial gibbsite, according to Rooksby [15]	
d, A	I	d, A	I	d, A	I	d, A	I	d, A	I	d, A	I
4.93	5	4.85	10	4.823	10					1.475	1
4.40	5	4.34	9	4.337	6	1.45	4	1.448	40	1.452	4
3.34	2—3	3.31	1	3.303	3	1.43	3			1.433	2
3.20	4	3.12	1	3.165	2	1.40	3	1.404	60	1.405	3
				3.086	1	1.39	2			1.394	2
2.46	3	2.45	7	2.451	5	1.37	1			1.375	1
2.37	4	2.38	7	2.374	5	1.36	2—3	1.352	20	1.356	3
		2.26	1	2.278	1	1.32	2			1.323	2
2.24	2			2.236	2	1.31	3	1.312	10	1.311	2
2.16	2	2.17	3	2.157	3					1.292	1
2.04	3	2.04	3	2.039	4					1.264	1
1.98	3	1.99	2	1.986	3	1.24	3			1.246	2
1.91	3	1.907	2	1.909	3					1.226	1
1.85	4									1.207	3
1.79	4	1.798	3	1.797	4	1.19	3			1.189	2
1.74	4	1.741	4	1.743	4	1.17	1			1.173	1
1.68	4	1.681	5	1.677	4	1.14	2			1.139	2
1.65	1	1.642		1.643	1	1.13	2				
				1.630	1	1.12	2			1.119	2
1.59	1	1.586	1	1.582	2					1.106	2
				1.569	1	1.08	2			1.089	1
1.56	1	—	—	1.548	1	1.02	3			1.077	1
				1.524	1						
				1.503	1						

Feodot'ev observed the same type of heating curves for both natural and artificial gibbsite. The latter was synthesized from chemically pure aluminum metal dissolved in caustic soda, with the concentrated solution of sodium aluminate obtained being subjected to carbonization. The fine white crystals of aluminum hydroxide precipitated in this way were filtered and used for seeding portions of the aluminate solution.

Aluminum hydroxide was separated from a subsequent portion of the aluminate solution by centrifuging. In this method the aluminate solution, heated to 45-50°C and seeded with a small amount of hydrate obtained by carbonization, was stirred for several days. A fine white powder of aluminum hydroxide trihydrate was precipitated, dried at 50°C, and subjected to microscopic and chemical analysis.

Although these artificial gibbsites were obtained under different conditions, the optical and thermal data are identical, whereas the structural and physiomechanical properties differ.

Gibbsite obtained by the electrochemical induration of clays is a dense crystalline material which serves as a firm cement for aggregates of clay grains; the monolithic body which it forms with the rock has a mono-axial compressive strength of 12 kg/cm², even when completely saturated with water.

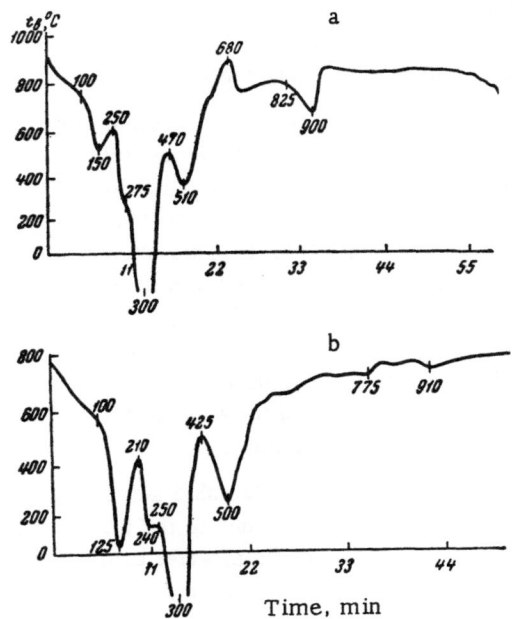

Fig. 9. Differential thermal curves: a) gibbsite with an admixture of montmorillonite; b) incompletely dressed gibbsite.

Belyankin et al. [16] discussed the double gibbsite effect on the heating curve. Dewatering experiments showed that thermal decomposition of gibbsite takes place in three stages, as follows:

1) $Al_2O_3 \cdot 3H_2O \rightarrow Al_2O_3 \cdot 2.5H_2O + 0.5H_2O$;
2) $Al_2O_3 \cdot 2.5H_2O \rightarrow Al_2O_3 \cdot H_2O + 1.5H_2O$;
3) $Al_2O_3 \cdot H_2O \rightarrow \gamma - Al_2O_3 + H_2O$.

The first endoeffect, due to the crystallization of artificial gibbsite, is absent in the specimen synthesized at normal temperatures, evidently because of the extreme degree of dispersion of the synthesized specimen, which prevents the appearance of the intermediate dehydration stage.

The conclusions in [16] confirm the heating curve of artificial gibbsite formed by electrochemical induration of clays; the crystal size of secondary gibbsite in this case is 0.02-0.001 mm, the disperse phase is absent, and therefore a distinct double endoeffect is obtained. However, if it is assumed that the first endoeffect occurs at a loss of 0.5 H_2O, it becomes difficult to explain the fact that "electrochemical" gibbsite possesses a double endoeffect and a formula $Al_2O_3 \cdot 2.47 H_2O$.

Since gibbsite appeared regardless of whether aluminum chloride was added to the electrolyte, and since sodium chloride was not used as an additive, it may be assumed that gibbsite formation takes place without intermediate sodium aluminates.

In an experiment which lasted 1800 hr involving aluminum electrodes and added aluminum chloride, the anode was intensely eaten away. However, no aluminum compounds had been formed at the anode by the end of the experiment, owing to the hydrogen-ion concentration (pH=3.3). Gibbsite and aluminum gel were formed, though, as a crust in the cathode zone, rising nearly 3-4 cm above the anode zone level. This experiment is proof of the active migration of aluminum from the anode to the cathode, apparently as a result of electro-osmosis (in which the electrolyte is transferred along with water). The pH of the medium is an important factor in this process.

But the formation of gibbsite by corrosion of the aluminum cathode evidently takes place partially by substitution of metallic aluminum, as may be clearly seen under the microscope. It can be seen that in some areas fragments of the electrode are replaced by gibbsite.

Experiments on Kynov argillite mud show that induration between structural units of the rock is related to gibbsite formation in the mud, confirmed by the thermal curves in Fig. 10 (endoeffect at about 300°C).

In other cases an interstructural cement bond of calcite and iron oxides is observed.

Allophane is the most widely occurring mineral of the allophanoid group and is formed in various clays during dc treatment with aluminum electrodes and a 0.1% solution of water glass. The presence of amorphous silica and free alumina in the initial rock also leads to the formation of allophane in a neutral or weakly alkaline medium (the allophane sometimes contains inclusions of pelitomorphic calcite, confirming the above).

Secondary allophane is white, frequently having a pinkish tint; the refractive index is about 1.477. The compositions of the individual allophane grains differ, since their properties vary within wide limits

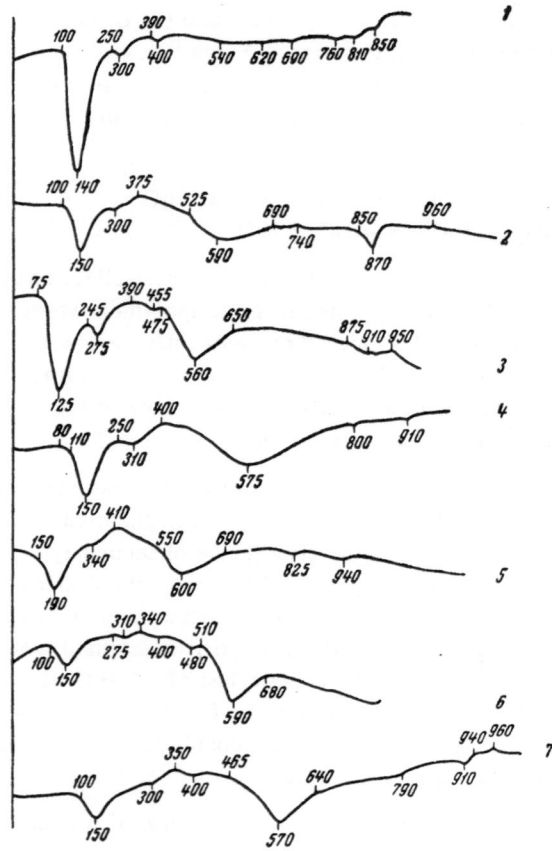

Fig. 10. Differential thermal curves of indurated Kynov clay; experiment 130: 1) cathode; 2) cathode 1st layer; 3) anode; experiment 131: 4) anode—cathode; 5) cathode—anode; experiment 132: 6) anode—cathode; 7) cathode—anode.

(the specific gravity, 1.8-2.17; the hardness 3-3.5). The allophane is isotropic and decomposes slowly in weak hydrochloric acid, forming a silica gel.

The chemical analysis of the dressed allophane is as follows (%): total SiO_2, 35.46; amorphous SiO_2, 33.7; TiO_2, 0.12; Al_2O_3, 24.70; Fe_2O_3, 2.58; CaO, 7.0; MgO, 2.55; Na_2O, 7.42; K_2O, 0.54; H_2O^-, 8.29; H_2O^+, 10.31; CO_2, 1.14; total 100.11. This composition corresponds to the following molecular ratios: 0.07 $\cdot Fe_2O_3 \cdot 0.17CaO \cdot 0.013MgO \cdot 0.5Na_2O \cdot 0.07K_2O \cdot 1.00Al_2O_3 \cdot 1.18SiO_2 \cdot 4.43H_2O$. The molecular ratio $SiO_2 : Al_2O_3 : H_2O$ is 1.18: 1.00: 4.43. The term "allophane" is usually employed for allophanoids with an $SiO_2 : Al_2O_3$ ratio of about unity.

The presence of a considerable admixture of Na_2O, K_2O, CaO, and MgO is due to the introduction into the indurated clay of stratal water containing a large number of Na, K, Ca^{2+}, and Mg^{2+} ions.

Although the investigations of Belyankin and Ivanova [17] dealt principally with the conversion of kaolinite during heating, they also touched on the genesis of the allophanoids. They showed convincingly that, like natural allophanoids, these allophanoids are obtained only by simultaneous precipitation of alumina and silica gels. With silica and free alumina present in these conditions, allophane is formed during the electrochemical induration of clays and clay rocks, the current acting as the reaction accelerator.

Thermal analysis of the allophane thus obtained showed an endoeffect at 50-200°C and a small exoeffect at 1000°C (Fig. 11), characteristic of allophanoids with different $SiO_2 : Al_2O_3$ molecular ratios; this agrees with the data of [18].

Belyankin and Feodot'ev [19] contrast the allophanoids with kaolin, as being collomorphic mineral masses with the general composition $Al_2O_3 : nSiO_2 \cdot pH_2O$, where $n \approx 1$ and $p \geq 4$. In its exothermic peak of 950-1000°C the allophanoid's characteristic curve is thermally similar to that of kaolin; it differs from the latter in the absence of both an endothermic peak at 550°C and a second exothermic break at ~1200°C.

X-ray structural analysis of synthetic allophane, obtained during the electrochemical treatment of clays, showed an admixture of a crystalline phase — possibly one of the forms of hydrated alumina similar to that obtained in [20].

X-ray structural analysis of this allophane was performed twice. Data on the interplanar spacings obtained in the first analysis are given in Table 6, those obtained in the repeat analysis (27 months later) in Table 7. The presence of a crystalline phase is evidently due to alumina; this supposition is confirmed by the more crystalline state of the allophane in the second analysis ([21] agrees with this to some extent).

According to [21], sols of hydrous alumina have the form of amorphous pellets with diameters up to 100 mμ. Crystallization commences 20-24 hr after formation of the sol. It consists 2-3 mo. later wholly of crystalline gibbsite. According to [21], in sols prepared at 80-90°C colloids of hydrous alumina possess a crystalline structure as soon as they are formed.

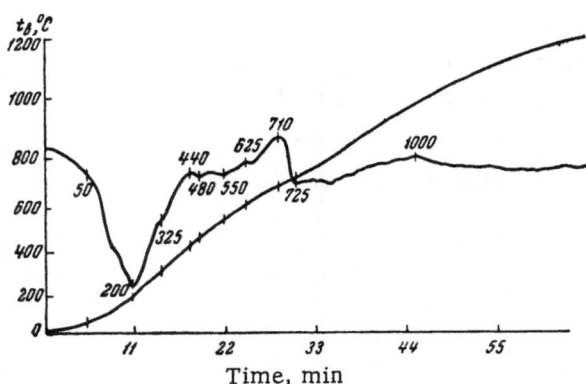

Fig. 11. Differential thermal curve of allophane.

Hisingerite is formed when a direct current acts on clay in the presence of an iron electrode and 0.1% water glass solution. It may also be formed without the latter, as a result of the iron and silica in the clay. Thus, we found that hisingerite was formed when a direct current acted on ferruginous montmorillonitic clay from the Ufa deposits (Bashkiriya).

Hisingerite is precipitated in both, an acid or neutral medium (Fig. 12), ferruginous dark varieties from the former, and lighter less ferruginous ones (often associated with allophane) from the latter.

Under the microscope, hisingerite is isotropic with an incrustation-collomorphic structure; the color ranges from straw yellow to dark brown. It is virtually opaque, depending on the ratio of silica to iron.

The refractive index of hisingerite varies widely (from 1.516 to 1.630 or more). The color intensity, refractive index, and specific gravity of hisingerite depend on the ratio of silica to iron.

Allophanoid, according to the authors, is a greenish mineral, free or almost free of silica, consisting mainly of hydrous alumina gels. Its refractive index is 1.582. Allophanoid is formed in the anode zone when a current is applied to Oglanly bentonite clay, using aluminum electrodes and added sodium chloride.

The refractive index of air-dried alumina gel containing about 35% moisture is 1.578-1.587 [22]. The ratio of Al_2O_3 to SiO_2 in allophanoid varies widely [23]. With a marked reduction in the moisture content allophanoids are converted to alumina hydrates, possibly of type similar to so-called "shanyavskite" ($Al_2O_3 \cdot 4H_2O$).

TABLE 6. Interplanar Spacings of the Crystalline Admixture in Allophane

Serial No.	I on the five-point system	d, A	Serial No.	I on the five-point system	d, A	Serial No.	I on the five-point system	d, A	Serial No.	I on the five-point system	d, A
1	3	6.65	13	2	1.68	7	5	2.98	19	1	1.32
2	3	5.77	14	3	1.60	8	4	2.70	20	1	1.28
3	4	5.12	15	1	1.54	9	4	2.42	21	1	1.21
4	4	4.16	16	2	1.49	10	4	2.12			
5	5	3.77	17	2	1.45	11	1	1.97			
6	5	3.31	18	3	1.37	12	3	1.74			

NOTE: Synthetic allophane: $CuK\alpha_{1,2}$ radiation; chamber diameter 57.3 mm; specimen diameter (column) 0.5 mm.

TABLE 7. Interplanar Spacings of the Crystalline Admixture in Allophane
(repeat analysis 27 months later)

Serial No.	Relative intensity I	Reflec- tion angle θ	Inter- planar spacing d, A	Serial No.	Relative intensity I	Reflec- tion angle θ	Inter- planar spacing d, A
1	3	6°09'	9.03	18	2	30°49'	1.89
2	2	7°45'	7.17	19	2	32°13'	1.81
3	4	8°39'	6.43	20	3	33°31'	1.75
4	4	11°19'	4.929	21	2	35°02'	1.69
5	3	12°34'	4.44	22	3	37°38'	1.58
6	2	13°34'	4.12	23	2	39°09'	1.53
7	5	15°10'	3.69	24	2	40°33'	1.49
8	2	16°52'	3.39	25	2	41°46'	1.45
9	2	17°40'	3.19	26	1	43°23'	1.41
10	3	18°28'	3.05	27	2	45°23'	1.36
11	4	21°41'	2.62	28	2	52°48'	1.21
12	3	23°51'	2.39	29	1	56°43'	1.16
13	1	24°47'	2.31	30	1	63°39'	1.08
14	4	27°18'	2.11	31	1	67°40'	1.05
15	2	27°56'	2.07	32	1	73°17'	1.01
16	2	29°07'	1.99	33	1	76°41'	0.99
17	3	30°19'	1.92				

NOTE: Synthetic allophane: $Fe\alpha_{1.2}$; chamber diameter 86 mm; specimen diameter (column) 0.5 mm.

Shanyavskite is a colloid mineral, forming greenish translucent crusts in the fissures of dolomites (Moscow region). The luster of shanyavskite is markedly hyaline or nacreous; its specific gravity is 2.3, and its hardness ≤ 3. The analytical data (in %) are as follows: MgO, 0.35; CaO, 2.28; Al_2O_3, 53.53; CO_2, 2.18; SiO_2, 1.33; H_2O, 40.95; total 100.62. The molecular ratio $Al_2O_3 : H_2O$ is 1.00:5.48 [23].

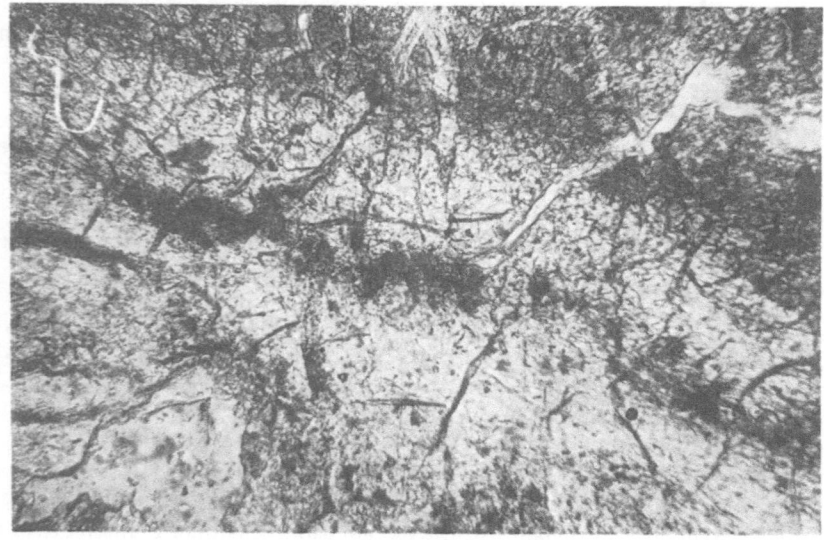

Fig. 12. Alternation of hisingerite (light) and aluminum gel (dark)
($\times$ 153, parallel nicols).

The shanyavskite from the Mineralogical Museum of the Academy of Sciences of the USSR (investigated by Chukhrov [24]) showed a gibbsite line on the powder patterns. Microscopic investigation showed that allophane is present, together with gibbsite, in shanyavskite aggregates.

In our investigations, the whole allophanoid group in the indurated clays was observed in the form of various types of cements.

A l u m i n i t e is formed in various clays when aluminum electrodes and a saturated solution of $CaSO_4$ are used, or if SO_4^{2-}-containing minerals are present in the clay (e.g., in the case of Jurassic clays from Podol'sk). The particular feature of these clays is the presence of jarosite with readily detachable SO_4^{2-} ions. Aluminite is precipitated as white needles which form curious star-shaped concretions growing in the indurated clay.

Under the microscope, it is seen that the aluminite crystals form long prisms or fibers with negative elongation and low birefringence, and direct extinction with $\gamma = 1.468$ and $\alpha = 1.460$. In the experiments aluminite is observed in paragenesis with calcite. In natural conditions (the Ermakov deposit on the Volga) aluminite is found in dolomite of the Kazan stage (Permian) [25].

Our data on synthetic and natural aluminite show that it is formed in an alkaline medium.

A l u m i n u m g e l is formed by corrosion of aluminum electrodes or by addition of aluminum chloride solution. The pH of the medium in which the gel is precipitated is 3.5-6.8; at other pH values the gel is in a dissolved state. Its refractive index varies over a wide range and depends on the degree of saturation by water. In the air-dry state the refractive index is 1.582. If the gel is dried at 105°C (to constant weight), it is converted to gibbsite.

O p a l i n e m a t e r i a l is formed during dc treatment if water glass is added to the clay, irrespective of the type of electrodes used. We have included this substance in opal by reason of its isotropicity and the refractive index n = 1.454. The hardness of the substance is 2.25, i.e., less than that for natural opal. The substance is found as an intimate intergrowth with pelitomorphic calcite, indicating simultaneous formation.

M a g n e t i t e is formed in the cathode zone in the presence of iron electrodes, or with an iron electrode as the anode and an aluminum electrode as cathode. A chloride must be added, because Cl serves as the gegenion for Fe^{3+}. With an Fe electrode as cathode and an Al electrode as anode, magnetite is not formed and iron is accumulated in the cathode zone. Magnetite is formed from iron hydrates.

During electrochemical induration of clays the location of the electrodes is of great importance for mineral formation. In work on mineral formation in a natural electric field [26], galena and molybdenite were used as electrodes. For formation of wulfenite and hydrocerussite, the electrodes were located as follows: anode — galena, cathode — molybdenite. With the reverse order (i.e., anode — molybdenite and cathode — galena), no dissolution of the minerals or electrodes was observed, and therefore secondary minerals could not be formed.

Magnetite is accumulated with montmorillonitic and monothermitic clays in an alkaline medium (Fig. 13), maximum accumulation being observed in experiments with iron electrodes and addition of ferrous chloride. It is precipitated in paragenesis with hydrogoethite (lepidocrocite). Magnetite is identified by the characteristic crystal shape under the microscope and by its magnetizability.

The magnetite obtained by induration was subjected to an X-ray investigation. Comparative data on the interplanar spacings of artificial and natural magnetite are given in Table 8. All the X-ray pattern lines agree closely with the tabular data for Fe_3O_4. Furthermore, the specimen contains a small admixture of montmorillonite, indicating agreement of the interplanar spacings of No. 2, 3, 5, and 10 with literature data for montmorillonite.

H e m a t i t e was formed in the cathode zone in experiments with montmorillonitic clay, using iron electrodes and adding ferrous chloride. The direct current time required for hematite formation was 200 hr; the current density was 39 ma/cm^2 up to 100 hr, and 15 ma/cm^2 up to 200 hr.

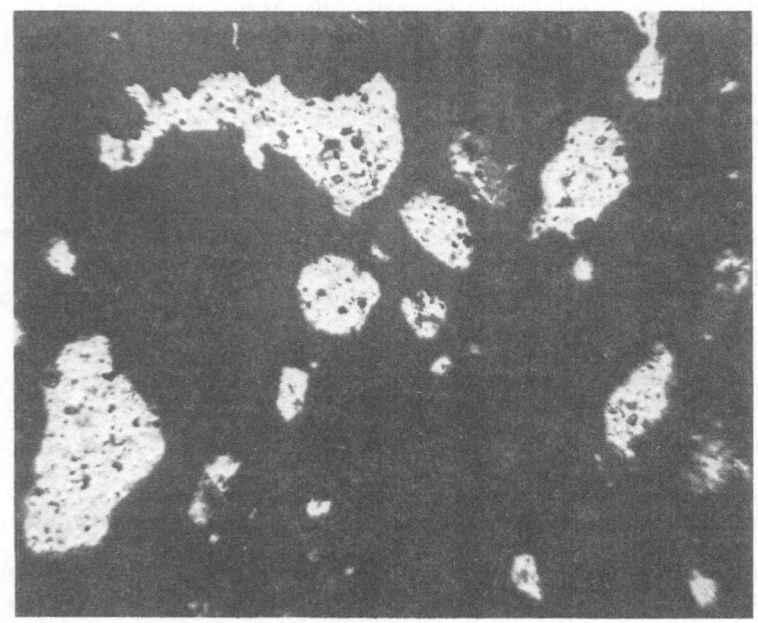

Fig. 13. Magnetite (dark) with clay (light) ($\times$ 153, parallel nicols).

The formation of hematite during induration is due to martitization, i.e., conversion of magnetite to hematite according to the equation $4Fe_3O_4 + O_2 = 6Fe_2O_3$.

TABLE 8. Interplanar Spacings of Synthetic and Natural Magnetite

Data on synthetic magnetite				Natural magnetite		Natural montmorillonite	
Serial No.	I on the 5-point system	θ	d, A	I on the 100-point system	d, A	I	d, A
1	1	11°31′	4.85	6	4.85	—	—
2	1	12°34′	4.41	—	—	Very strong	4.45
3	1	18°44′	3.01	—	—	The same	3.03
4	3	19°12′	2.96	28	2.97	—	—
5	5	22°32′	2.52	100	2.53	Very strong	2.53 *
6	1	23°35′	2.42	11	2.42	—	—
7	3	27°32′	2.09	32	2.10	—	—
8	2	34°28′	1.71	16	1.71	—	—
9	3	36°58′	1.601	64	1.610	—	—
10	4	40°50′	1.48	80	1.48	Very strong	1.49 *
11	1	46°52′	1.325	6	1.326	—	—
12	2	49°16′	1.276	20	1.279	—	—
13	1	53°20′	1.205	5	1.210	—	—
14	1	60°04′	1.120	10	1.121	—	—
15	3	62°48′	1.089	32	1.092	—	—
16	2	67°40′	1.046	10	1.049	—	—

*Fe_2O_3 + montmorillonite.

During the experiments there was a steady change in the redox potential and pH of the liquid phase, so that magnetite was partially converted to hematite by the time induration had proceeded for 200 hr. At the end of the experiment (1800 hr) the cathode had an acid medium (pH = 5.5) as a result of gradual oxidation of the indurated rock.

Hydrohematite is formed in clay rocks by direct current in a neutral or weakly acid medium, as a result of corrosion of the iron electrode. Hydrohematite is red to dark red, anisotropic, fine-crystalline, and frequently collomorphic. In many cases it is found in association with hisingerite and limonite.

Hydrogoethite (hydrolepidocrocite) is formed in the cathode zone when iron electrodes are used, or when ferrous chloride is added and graphite electrodes employed.

In [27] it was established that goethite and lepidocrocite are formed at pH = 11.

Hydrogoethite and hydrolepidocrocite are fine-grained and fine-fibered pleochroic aggregates.

In hydrolepidocrocite the pleochroism is more distinct and its tint lighter.

Limonite is formed when a direct current is applied to any clay, using iron electrodes and adding sodium, in an acid medium near the anode electrode. It strengthens the rock, forming veinlets within it and filling cavities. In ordinary conditions ferric hydroxide in the form of limonite is precipitated at pH = 2; in the experimental conditions (with chlorides present) the pH is higher. The chemical composition of the limonite selected by hand is as follows (in %): SiO_2, 12.72; Al_2O_3, 3.83; Fe_2O_3, 65.30; CaO, 0.40; MgO, 0.99; $H_2O^- + H_2O^+$ = 16; total 99.24, corresponding to the formula $Fe_2O_3 \cdot 2.17H_2O$. A characteristic feature is that the silica and aluminum found during analysis are present here in the same ratios as in the initial montmorillonite used for the experiments — evidently of inadequate purity when sampled. This supposition agrees with the results of thermal analysis, which shows characteristic breaks on the heating curve.

X-ray structural analysis of synthetic limonite showed that most of the strong lines, and some of the weak ones, coincide with lines characteristic of limonite (goethite). Comparative data on synthetic limonite with natural limonite and goethite are shown in Table 9.

Nontronite is formed in various montmorillonitic clays when iron electrodes are used, as a result of partial exchange of an aluminum cation of the clay by an iron cation from the electrode.

Under the microscope nontronite is seen as green or pistachio-green platelets with medium birefringence. The refractive indices are $\gamma > 1.585$ and $\alpha < 1.585$. Nontronite formed from sodium montmorillonite had lower indices: $\gamma >$ and $\alpha < 1.540$.

Calcite is deposited near the cathode; its formation is due to calcium salts in the aqueous solution saturating the clay, and to calcite transferred from other parts of the rock. Near the anode there is usually intense dissolution of calcite (if present in the clay), and it is transferred to the cathode where the clay is enriched with calcite. Maximum accumulation takes place with bentonite clay.

Observations under the microscope show that calcite is most usually found as pelitomorphic grains with o = 1.654 and ϵ = 1.481; idiomorphic macrocrystals are formed less frequently during induration of clay for 100 hr.

Magnesite was formed in bentonite clay (Oglanly) on addition of $MgCl_2$ and a 0.1% solution of water glass (current time 65 hr), using aluminum and iron electrodes. It is deposited in an alkaline medium. Under the microscope magnesite is seen as aggregates of fine rhombohedral crystals with well-formed faces.

The thermogram of the magnesite obtained in this way shows the break characteristic of natural magnesite. An association of magnesite with opaline material, formed as a result of the addition of water glass, is observed.

No data on magnesite formation are as yet to be found in the literature; there is but a reference to the formation of synthetic hydromagnesite ($MgCO_3 \cdot 3H_2O$).

TABLE 9. X-Ray Structure Data of Synthetic and Natural Limonite

Investigated specimen		Limonite from literature data (VIMS)		Goethite (Kitaigorodskii)		Investigated specimen		Limonite from literature data (VIMS)		Goethite (Kitaigorodskii)	
d, A	I	d_{sep} KX	I	d, A	I	d, A	I	d_{sep} KX	I	d, A	I
		4.96	3	4.98	4	1.60	3	1.694	1	1.60	8
4.22	4	4.15	10	4.21	100	1.56	3	1.554	4	1.56	28
3.35	4	3.363	3	3.39	12	1.50	5	1.500	3	1.50	29
2.96	4					1.47	4	1.465	1		
2.71	4	2.674	6	2.70	36			1.459	1		
2.54	5	2.565	3	2.58	24	1.45	1	1.445	3	1.455	12
		2.508	1					1.412	2	1.420	4
		2.471	3					1.386	2		
2.43	5	2.433	7	2.45	80	1.36	1	1.361	1	1.355	8
2.27	1	2.237	4	2.25	12			1.351	2		
2.17	1	2.175	4	2.19	20			1.340	1		
2.08	2	1.997	1	1.92	8	1.31	2	1.309	2	1.315	12
		1.908	2			1.27	1				
1.79	1	1.788	2	1.80	8	1.19	1				
		1.762	1			1.18	1				
1.71	4	1.709	6	1.72	36	1.14	2				
		1.679	3	1.68	4						
		1.650	1	1.65	4						

Gypsum was formed by the action of direct current on rock containing jarosite and calcite or Ca^{2+} and SO_4^{2-} ions in the aqueous solution used for wetting. This secondary gypsum consists of colorless prisms or fibers with oblique extinction, low birefringence and refractive index $\gamma = 1.530$ and $\alpha = 1.522$.

Solntsev and Borkov [10] attempted to strengthen the middle soil by forming an insoluble residue in the form of $CaSO_4 \cdot 2H_2O$, assuming that gypsum is formed when Ca^{2+} ions from the anode meet SO_4^{2-} ions from the cathode. According to our data, gypsum is precipitated from the solution on to the surface of the indurated rock; it does not cement the rock, being formed on its surface and thus not bonding the rock into a monolithic structure.

THE EFFECT OF THE ELECTRODE MATERIAL ON SECONDARY MINERAL FORMATION AND THE NATURE OF THE INDURATED CLAY

The composition of the secondary minerals normally includes material from the electrodes, the electrolyte additive, and elements entering the clay as a result of redox reactions in the clay–electrolyte–water system. The cathode zone is mainly impregnated with Na^+, Ca^{2+}, Mg^{2+}, Fe^{2+}, Al^{3+}, H^+, and other cations, and the anode zone with HCO_3^-, SO_4^{2-}, Al^{3-}, Fe^{3-}, OH^-, O^{2-}, and other anions.

By accumulating and crystallizing in the aqueous medium between the grains of the clay aggregates, the secondary minerals strengthen the bonds between them and cement the clay, thereby increasing its strength. The fissures and pores of the rock are occupied by an artificial mineral framework (cement) and the newly formed mineral is accumulated in the form of pure crusts 1 cm thick on the surface of the rock beneath the electrodes.

Vigorous dissolution and transfer of iron and aluminum (electrode material) hydrates is observed in an acid medium; formation of new minerals takes place primarily in an alkaline medium.

The induration obtained with carbon electrodes and aluminum and iron chloride solutions is inferior to that obtained in the same conditions but without these solutions.

From [28] it is known that in natural conditions iron and alumina hydrates are mobile in an acid medium and in the dissolved state (as sulfates). In the induration experiments sulfuric acid was not used as an additive, so that it is highly improbable that movement of iron and alumina as sulfates occurred. The most usual additives were NaCl, $CaCl_2$, $MgCl_2$, etc., and it may therefore be assumed that iron is transferred as a chloride.

Various new minerals are accumulated in the cathode and anode zones. The following minerals are formed in the cathode zone: magnetite, Ca, Mg, and Na carbonates, gibbsite, hydrogoethite, lepidocrocite, and hematite. Limonite and nontronite are formed in the anode zone; if the medium is neutral, gypsum, hydrohematite, the allophanoid group, aluminite, and opaline material are also formed (depending on the experimental conditions).

We must now examine the conditions of formation of hydrated iron and mixed iron oxide in the cathode and anode zones.

Maximum accumulation of iron in the form of magnetite, hydrogoethite (lepidocrocite), and hematite in the cathode zone is due both to addition of chlorides and to the presence of the iron electrodes. Under these conditions the iron hydrates are positively charged and accompanied by a specific micelle composition $[mFe(OH)_3 + nFeOCl + FeO]^+Cl^-$. When iron is precipitated in an acid medium, i.e., at the anode in experiments with iron electrodes and alternating addition of ferrous chloride and NaOH, the micelle composition changes to $[mFe(OH)_3 + nNaOH + OH]^-Na^+$. The cation is Na^+ in the anion complex. These phenomena observed in our experiments confirm the earlier conclusions of Chukhrov [24].

However, the results of a large number of experiments make it reasonably certain that the behavior of hydrated iron during induration is exactly the same as that of hydrated alumina. Experiments carried out with aluminum electrodes and additions of chlorides showed accumulation of aluminum hydroxide as gibbsite in the cathode zone at pH 10-12. In the anode zone in the pH range 3.5-6.8, an aluminum gel mixed with bentonite clay is formed (as shown by the refractive index). In experiments using aluminum electrodes and additions of NaCl, aluminum is accumulated in the cathode zone as gibbsite, and in the anode zone as siliceous allophanoid.

It may be concluded that, depending on the conditions, aluminum, like iron, can either acquire a negative charge and move toward the anode, or a positive charge, with movement toward the cathode. In such cases the cathode and anode zones are strengthened to the same extent.

THE EFFECT OF ADDITIONS OF STRATAL WATER, ELECTROLYTES, SALTS, AND A 0.1% SOLUTION OF LIQUID GLASS ON CLAY INDURATION

Additives affect clay induration in two ways: first, addition of chlorides accelerates electrode corrosion and is associated with movement of iron; the second effect is the formation of secondary minerals. For example, magnesite is formed when $MgCl_2$ is added to montmorillonitic clay, magnetite, and hydrogoethite (lepidocrocite), and hydrohematite and limonite are formed as a result of the dissolution of iron electrodes or by addition of $FeCl_3$.

Gibbsite is deposited when aluminum electrodes are used, with or without the addition of $AlCl_3$, in any clays in which conditions are suitable (i.e., an alkaline pH). Opaline material is formed by electrolysis of water glass, but the allophanoid group is formed when alumina hydrate or free hydrated iron is added to the water glass.

Laboratory experiments simulate natural conditions where the coagulators of colloidal hydrated aluminum are various electrolytes or oppositely charged colloids (e.g. silica, etc.). As a result of coagulation of the oppositely charged particles of hydrated alumina, aluminosilica gels of allophanoid appear. Aqueous gels of iron and of silica are coagulated in the same way, forming hisingerite.

Aluminite was formed in one case as a result of aluminum electrode corrosion and the SO_4^{2-} ions liberated from the jarosite of Jurassic clay, and in another case by addition of a saturated $CaSO_4$ solution and aluminum electrode corrosion.

THE EFFECT OF ADDITIONS OF STRATAL WATER, ELECTROLYTES, SALTS, AND A DISPERSOLUTION OF LIQUID GLASS ON CLAY INDURATION

THE EFFECT OF CURRENT TIME ON THE SECONDARY
MINERAL CHARACTER

The experiments showed that secondary mineral character depends on the duration of the current. Limonite, gibbsite monohydrate, and hydrohematite are formed after an induration time of 5 hr, pelitomorphic calcite after 15 hr, opaline material 20 hr, gibbsite trihydrate and hisingerite 30 hr, magnesite 65 hr, and hematite after more than 100 hr. The duration of an experiment also affects the shape and crystallinity of the secondary minerals (such as gibbsite, calcite, magnesite, magnetite, and others).

Experiments with shorter current durations showed that calcite is deposited as a pelitomorphic aggregate with a hardness of 2-2.5 and is principally a pore filler; however, in the course of time (observed to date only after 100 hr) idiomorphic calcite is formed. This calcite's firm cementation of the rock markedly strengthens the bonds between the aggregates of clay grains.

With an induration period of more than 30 hr, gibbsite forms large crystals of the order of ~ 0.01 mm (up to 30 hr it is deposited as ~ 0.001-mm crystals). After about 100 hr, gibbsite is often found as an overgrowing cement (crystal size ~ 0.01 mm) firmly bonded to the rock, completely filling the pores and converting the rock to strong monolith. These phenomena are observed when aluminum electrodes are used, but disappear when aluminum chlorides are added. Analogous phenomena are observed when aluminum electrodes are used and water glass added, the latter accelerating gibbsite crystallization.

The changes in allophane occur not only during the experiment, but the crystals formed in it continue to grow after removal of the electric field if the mineral is kept for a sufficient period of time. X-ray structural analysis revealed weak lines on the powder pattern (the specimen was left for two years after induration), which we attribute to formation of alumina, since chemical analysis showed the allophane to contain amorphous silica. Another X-ray analysis was performed twenty-seven months after the first. The crystalline lines on the powder pattern were more distinct, the particle size being ~ 0.0001 mm. The crystallization forces evidently continue to act on the rock, thereby strengthening it.

RELATIONSHIP BETWEEN SECONDARY MINERAL ACCUMULATION AND CURRENT DIRECTION AND STRENGTH

Secondary mineral accumulation in the anode or cathode zones depends on current direction. The current strength also has an effect on the degree of formation of individual secondary minerals. The depth of electrolyte penetration depends on the current period, direction, and strength, and on the added electrolyte.

The action of a direct current on wet clay intensifies the processes — mainly chemical reactions — taking place within it, owing mainly to reaction between the electrodes and electrolytes, and causes intense formation of secondary minerals.

THE EFFECT OF THE COMPOSITION AND pH OF THE LIQUID PHASE ON SECONDARY MINERAL COMPOSITION AND PROPERTIES

The pH of the medium and the redox potential determine what forms of iron and aluminum minerals are deposited, and the conditions of accumulation or solution of the individual minerals. Interesting data were obtained in protracted experiments with a variable pH at the cathode. The experiments were carried out on bentonite clay using aluminum electrodes and adding $AlCl_3$. Up to a current time of 200 hr, gibbsite was formed in the cathode zone, and the clay had a punch strength of > 21 kg/cm^2; after a period of 1800 hr, half the gibbsite had dissolved; the remainder had a granular structure of reduced strength and in many cases the gibbsite grains were covered with an aluminum gel. All these facts confirm the presence of an acid medium in the cathode (the pH of the aqueous extract is 5).

CLASSIFICATION OF ELECTROCHEMICALLY INDURATED ROCKS ACCORDING TO STRUCTURAL AND TEXTURAL CHARACTERISTICS

The structural characteristics of the pre-experimental and the indurated rocks were studied in thin sections. The properties of the rock, the material and petrographic properties, size and shape of the component particles, and their mutual location, determine the "rock structure" (or "texture").

The term "rock structure" indicates those elements which determine the shape and size of the component parts; "texture" has to do with those which determine their distribution.

The aim of our work was to impart stability to borehole wall rocks which become wet and swell; it was similar to the "profile" of engineering geology which, by studying the rock, also establishes a classification based on the laws of rock formation and genesis. In engineering geology, rock classification is based on the decisive characteristics determining rock behavior beneath a building.

The two principal divisions of classification of rocks are the composition of rock material and the composition of the cement. Consideration must be given to the formation of structural bonds in rocks, increasing or reducing their strength, and the possibility of deformation and permeability of water. The term "rock strength" denotes the capacity of the rock to withstand a given load (in kg/cm^2) during monoaxial compression tests.

However, the strength must be considered in conjunction with the structural and textural characteristics and rock composition, and the amount and type of synthetic cement. The type of cement and its location determine the rock strength, where the rock's bearing capacity and load-bearing section are sharply defined.

If the rock contains at least 40% basal cement, the basic strength is determined by the cement's mineralogy; fragments cemented in the rock play a subordinate role. In the case where the film, pore, and film-pore cement does not exceed 15%, the basic strengthening is related to fragments in the rock. During monoaxial compression of these rocks the load is borne by rock fragments in contact with one another, and the rock's strength will be characterized by the fragments' mineralogical composition.

In our experiments on the production of synthetic rocks, only one texture was predominant - the disordered one, where the aggregate of clay particles and cement are located without any order or orientation. Nevertheless, we obtained a great variety of structure, and from the combination of the groundmass of clay and synthetic cement we established the presence of a number of artificial structures — fragmental, conglomerate-like, pelitomorphic, collomorphic, and a substitution structure [29].

1. The fragmental structure is consertal, and the rock fragments or lumps have an isometric shape (Fig. 14). The cements of this rock are the artificial minerals gibbsite, calcite, allophane, etc. These

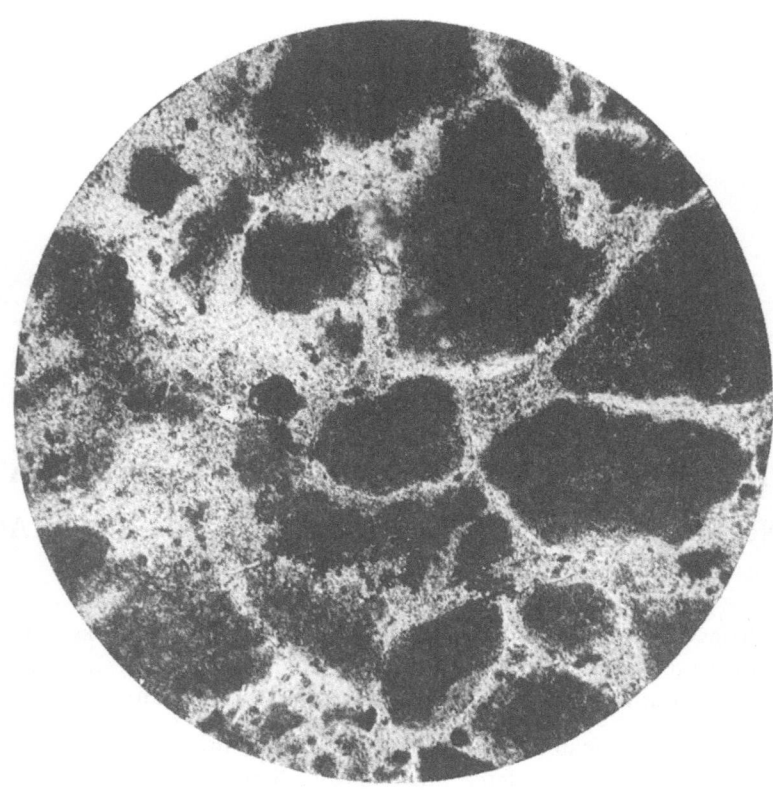

Fig. 14. Fragmental structure (× 90, parallel nicols).

fragments consist mainly of aggregates of clay particles and minerals, which form an admixture in the clay rock. The amount of cement differs in different parts of the rock. In some parts of the first layer (from 0-20 mm) it may form 50-60% — it sometimes forms a continuous monomineralic crust (depending on the induration time) with a thickness ranging from a few millimeters to 1 cm or more. But another factor (apart from the layer) determining the amount of cement accumulated is distance from the electrode: the greater this distance the smaller is the accumulation. In the second layer (from 20-40 mm) the cement is only 5-10%.

2. Conglomerate-like structure is represented by isometric clay fragments and mineralized fragments of the aluminum electrodes. The rock cement is mainly gibbsite, allophane, and calcite. The mineralized fragments of the aluminum electrode, with the initial mineralization, form a finely dispersed mass (Fig. 15) consisting apparently of a mixture of gibbsite and the remains of the aluminum electrode (gibbsite monohydrate). Monomineralic, very coarse gibbsite is formed at the peripheral part of the fragment. But gibbsite formation in larger crystals is due to the greater access of water and the lower packing density of the clay aggregates.

3. The pelitic structure consists mainly of aggregates of clay particles; the cement in this structure is allophane and finely dispersed artificial calcite (Fig. 16). The pelitic structure is the product of induration of carbonate-montmorillonitic clay. Some areas frequently contain idiomorphic calcite formed by recrystallization of pelitomorphic calcite deposited in the cathode zone. Crystal growth capacity and distribution of the crystallization centers are of great importance in crystallization of colloids and recrystallization of monomineralic crystalline aggregates. The formation of grains with a high growth capacity leads to well-developed crystals, as observed in indurated rock.

Radiate-fibrous aggregates are frequently formed from simple gels which yield monomineralic masses on crystallization. The crystallization nuclei were evidently formed either at the center of mineral formation or on the surface of a foreign inclusion.

48

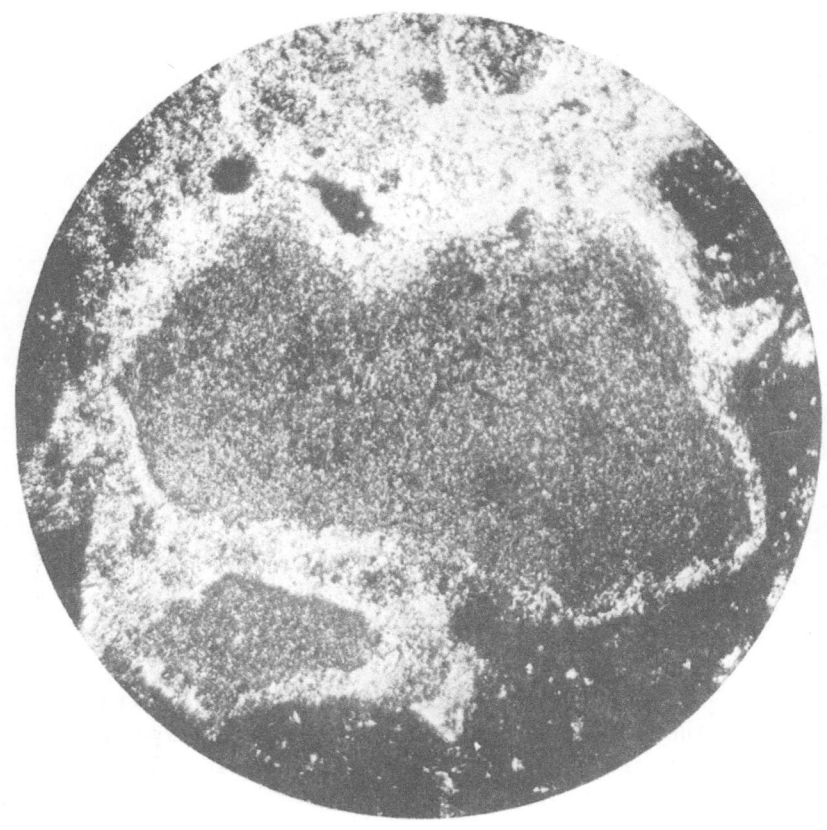

Fig. 15. Conglomerate-like structure (× 100, crossed nicols).

On the crystallization of colloidal formations and recrystallization of formations of complex mineralogical composition, crystalline grains of minerals will be formed, depending on their crystallization growth capacity. Fine and cryptocrystalline formations are the most characteristic type of concretion of recrystallized aggregates; this phenomenon is called collective crystallization, of which idiomorphic synthetic calcite formed after 100-hr electrochemical induration is an example.

4. The incrustation-collomorphic structure is represented by monomineralic formations of hisingerite and hydrohematite, the latter lying as oval or loop-shaped incrustations. Hisingerite is formed from any rock which contains free silica and iron (the materials from which it is formed). Hydrohematite is formed in any rock in a neutral or weakly acid medium. Free iron from any source is required for its formation (from dissolution of the iron electrodes, addition of ferric chloride and, finally, from the rock itself).

By the "structure" of ore minerals (which include hydrohematite and hisingerite) is understood the concretion of crystalline grains in shape, size, and location. The crystalline grain of the mineral is the morphological unit of structure.

5. Substitution structure is mainly secondary gibbsite, replacing the initial clay rock. Substitution by gibbsite is observed in some places, but clay remains only as individual remnants or islets. However, the chemical process takes place unilaterally; the clay rock itself is not dissolved but converted to another form; this procedure is typified by aggregation. Induration of the rock is due to accumulation and crystallization of secondary minerals. As a result of accumulation the original rock is displaced into the inner layers, i.e., gibbsite first cements the fissures, pores, and cavities; at the edges of the latter crystallization centers appear,

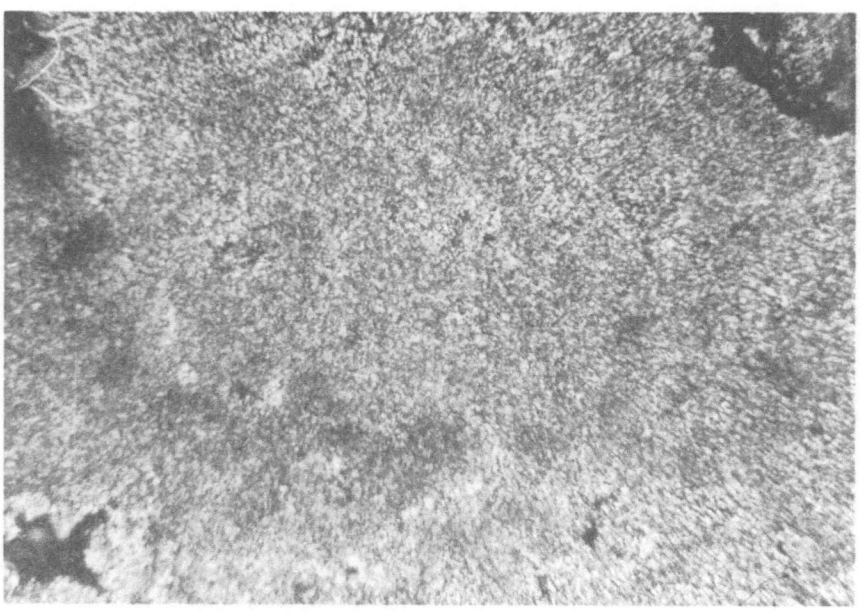

Fig. 16. Pelitomorphic calcite (× 340, crossed nicols).

directed toward the hollow part, pores, and cavities. After saturation of the inner layer, gibbsite is concentrated on the clay surface. This structure is a very strong one; the gibbsite crystals are formed without any orientation and therefore do not possess slip planes. Specimens of indurated clays with this structure showed the best results when subjected to monoaxial compression.

FORMATION OF SYNTHETIC CEMENTS DURING ELECTROCHEMICAL INDURATION OF CLAYS AND CLAY ROCKS

In natural conditions the solutions impregnating the pores of the deposit or rock are an important cementation agent. Under the effect of a change in pH, temperature, pressure, or composition of the dissolved salts and gases, or on reaction with minerals or organisms in the deposit, the dissolved substances are precipitated and occupy the interstices between the grains, thereby cementing them.

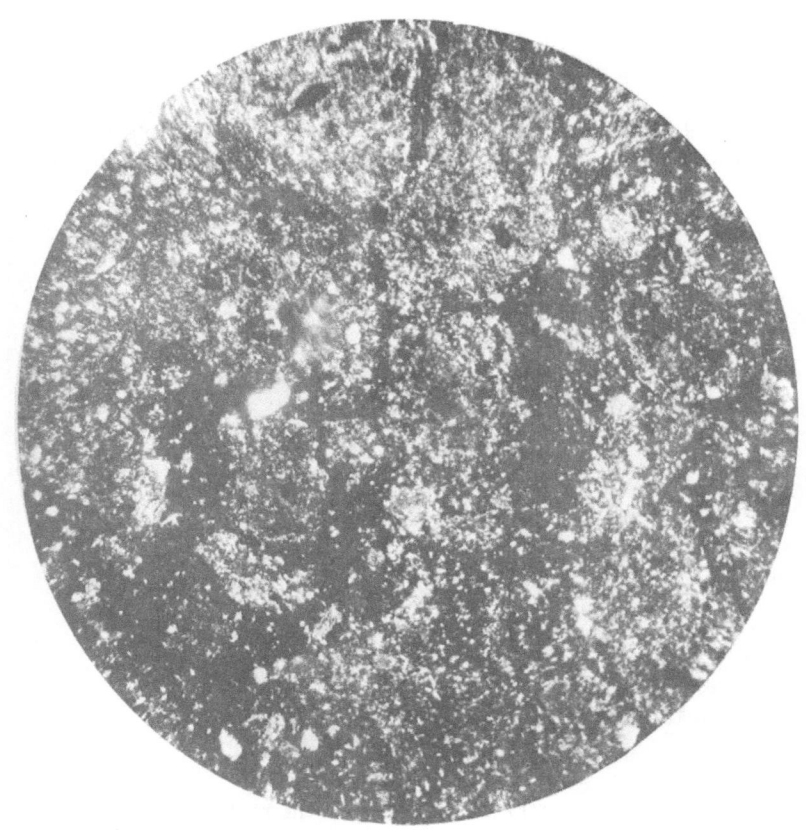

Fig. 17. Cement of the pores (black) occupies pores in light-gray rock (× 90, crossed nicols).

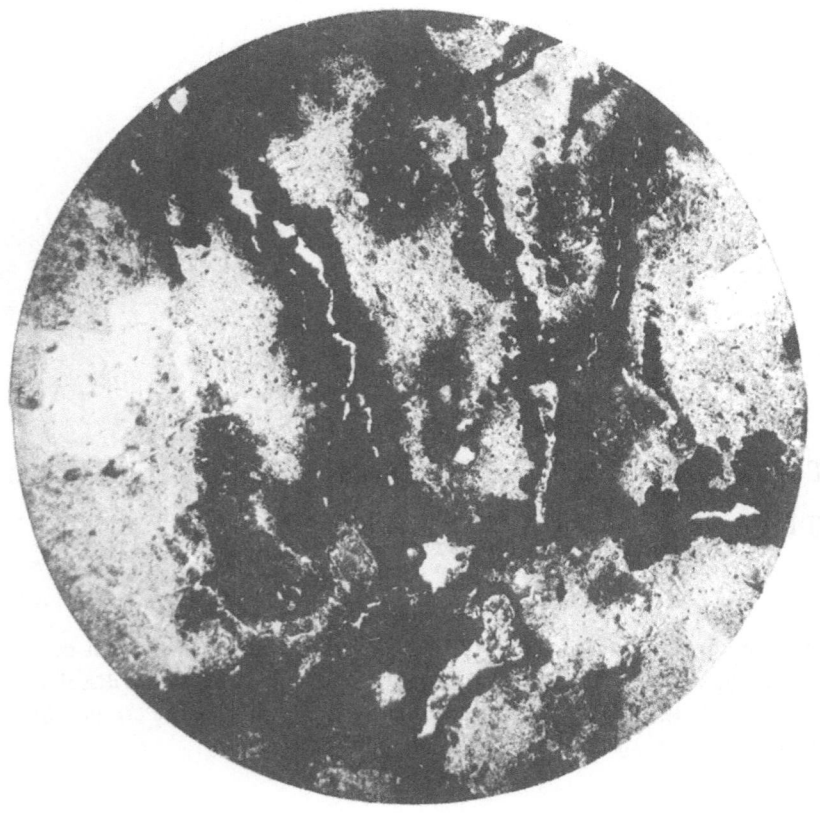

Fig. 18. Allophane (white) occupies cavities and pores in (black) limonite, cementing the rock ($\times$ 90, parallel nicols).

Cement penetration during clay induration is a very complex process. Cementation of clay and clay rocks depends on the composition and penetration of the electrolyte, the temperature and pH of the liquid

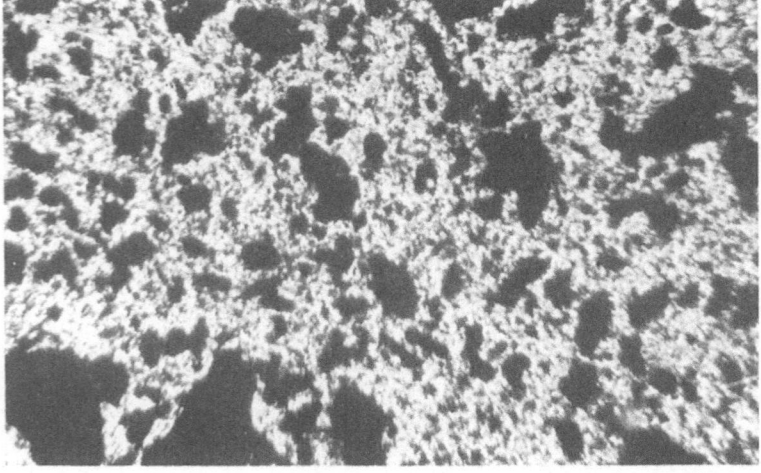

Fig. 19. Cluster, mottled, limonite cement (black) cements the rock (light) ($\times$ 153, parallel nicols).

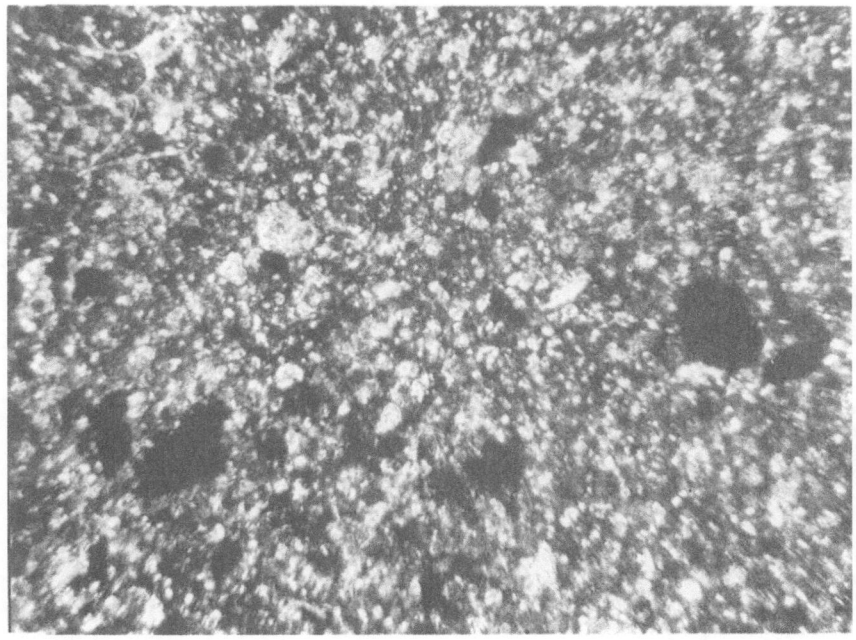

Fig. 20. The cement consists of calcite crystals (white) (× 340, crossed nicols).

phase, and the current strength and direction — i.e., on the experimental technology. Intense cement formation is observed in concentrated solutions; silt is deposited in the pores and the subsequent chemical reactions
lead to crystallization processes.

Cements of different types are formed at different distances from the electrodes. We investigated cements in transparent sections, in the preparation of which account was taken of the direction of the electric
field and the position of the electrode plates (perpendicular). The upper layers of the rocks have primarily
a basal cement, the lower layers mainly a film, pore, or cluster structure. Different cement types may therefore be encountered in the same thin section of indurated rock. Two different cements are sometimes observed, the second one more usually in the same form as the first; the former is formed by a change in the
method of introducing the additives, or on the use of new electrodes.

Different types of cement textures can be distinguished from the reaction between aggregates of clay
particles and cement particles in the indurated rocks.

The greatest amount of cement is observed in the cathode zone where optimum induration occurs; lesser
amounts of cement are found in the central, and even less in the anode zone. In the latter the rocks are only
strengthened if a particular technology is employed, with deposition of negative iron as limonite and without
siliceous allophanoid or markedly ferruginous hisingerite. In other cases induration does not take place in the
anode, but the particles aggregate owing to physical and chemical adsorption. The cements in the indurated
rock are primarily monomineralic, being represented by gibbsite, calcite, allophane and allophanoid, hisingerite and iron hydroxides (limonite), hydrogoethite (lepidocrocite and hydrohematite), and magnetite. The
mineral cements formed during electrochemical induration (hisingerite, limonite, and hydrohematite) sometimes yield a collomorphic-incrustation structure.

Basal cement in indurated rocks is a fine-grained or amorphous mass in which the clay grains are immersed but not in contact. This type of indurated rock cementation is very strong and characterized by the
strength properties of the cement. In indurated clays and clay rocks, the basal type of cement is formed near
the induration surface and directly upon it. Sometimes the cement forms a monomineralic film on the mineral's surface. The basal cement consists mainly of allophane, gibbsite, calcite, and lesser amounts of other
minerals.

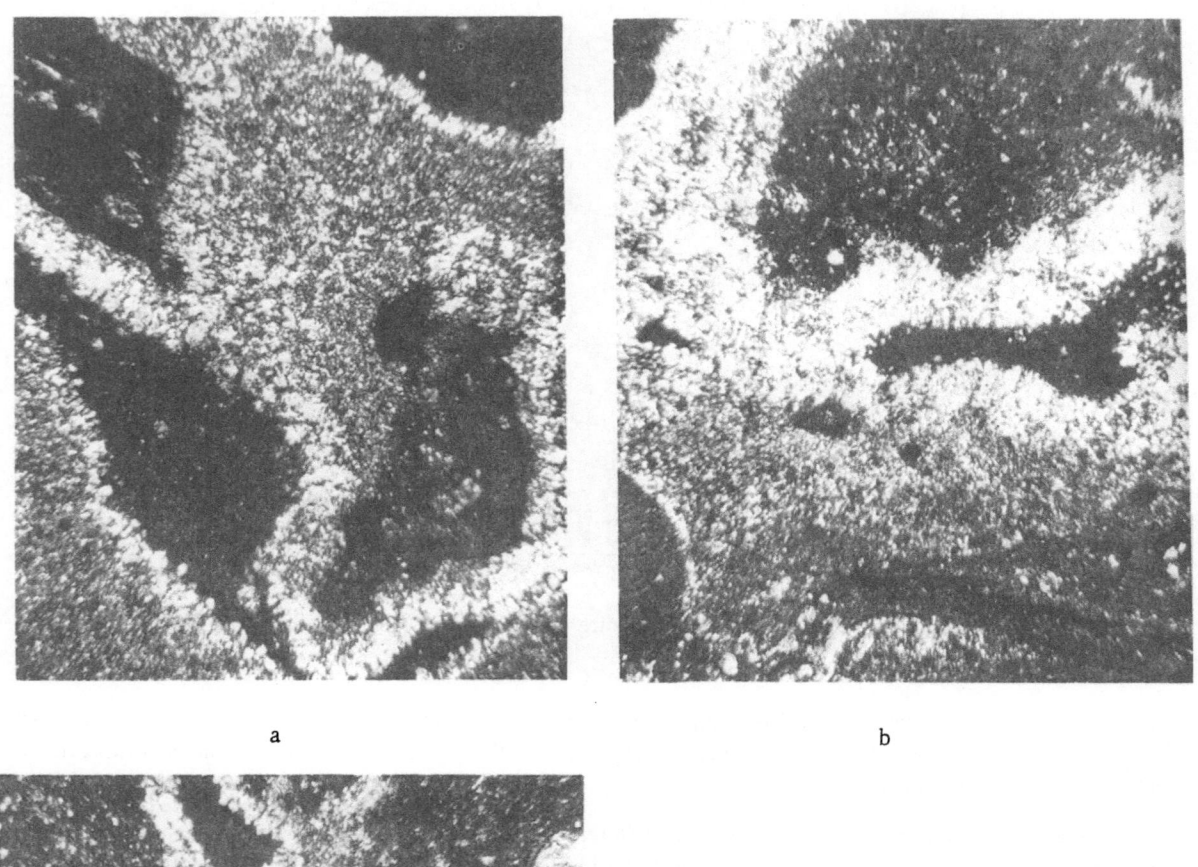

a

b

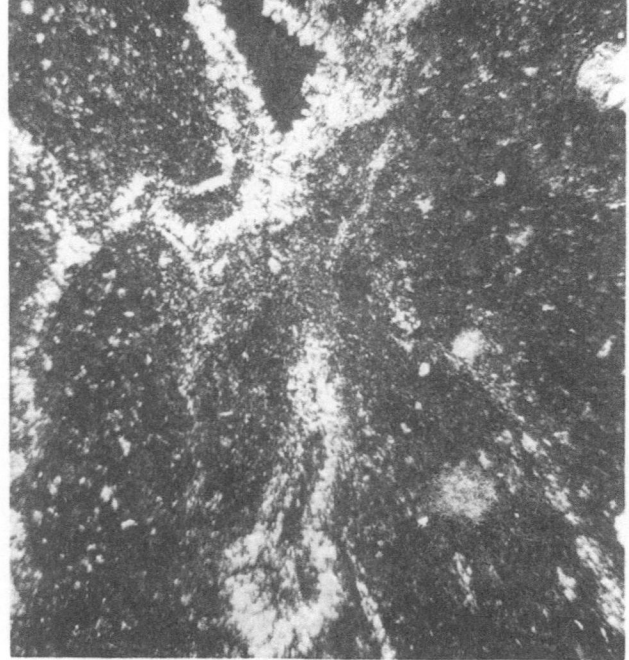

c

Fig. 21. Gibbsite overgrowth cement (white)
(× 153, crossed nicols).

Pore cement fills (to a greater or lesser extent) the area between the aggregates of clay particles in the indurated rocks, and is formed at a depth of at least 20-40 mm (Fig. 17). The cement's strength varies.

Filling cement is formed in or occupies cavities (pores) left between clay particle aggregates (Fig. 18), cemented by different type of cement formed during electrochemical induration. Filling cement is an allophanic, primary synthetic cement consisting of iron hydroxide.

Cluster or mottled cement is distinguished by its irregular distribution in the indurated rock, and consists of hydrated iron and calcite (Fig. 19).

Irregular-granular cement consists of an accumulation of individual idiomorphic crystals of calcite, sometimes of different size (Fig. 20); idiomorphic calcite is a strong cement in indurated rocks.

Overgrowth cement is represented by small cement crystals on the walls of pores and fissures. Growth of the crystals is toward the pore center, and they are elongated in this direction (Fig. 21). Overgrowth cement in indurated rocks is represented by gibbsite, the crystals being the largest of any obtained during electrochemical induration.

Mixed Types of Cements. These are cements of different composition and structure, represented by the following modifications: allophane—limonite (Figs. 22-23), allophane—hisingerite, allophane—calcite, limonite—hisingerite, hydrohematite—hisingerite, gibbsite—calcite, gibbsite—allophane, etc. The cement may have several different structures in the same specimen (basal, mottled-cluster, pore, and film); these differ according to depth of induration of the rocks, and also laterally (depending on the electrode distance).

According to the technological conditions, duplicate cements are observed (sometimes in the same form), and in some cases basal cement alternates with other types. The material composition of the latter may be the same, but the crystallinity differs: crystalline cement is replaced by amorphous cement, and vice versa. The mixed type is the predominant cement in indurated rocks.

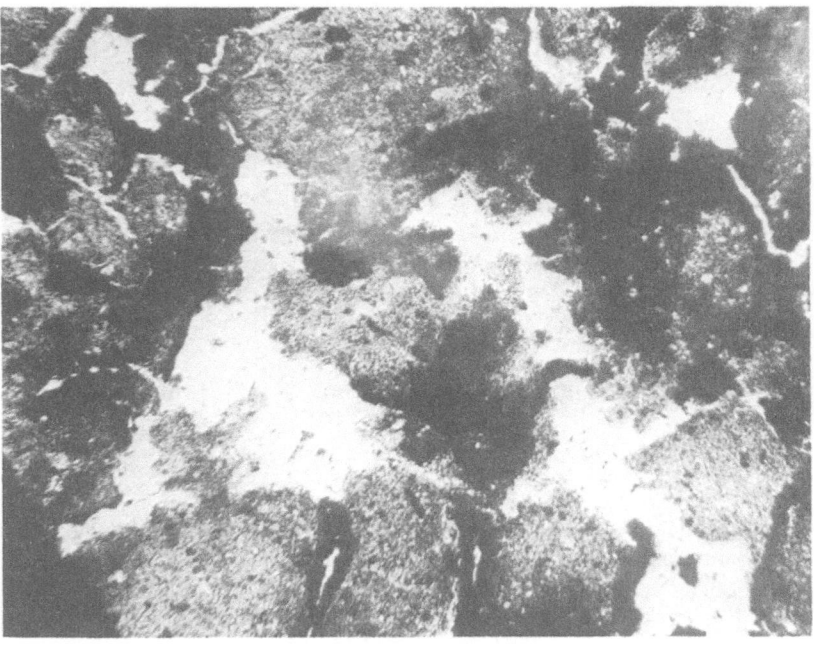

Fig. 22. Limonite (black) and allophane (white) cement the rock (gray)
(× 340, parallel nicols).

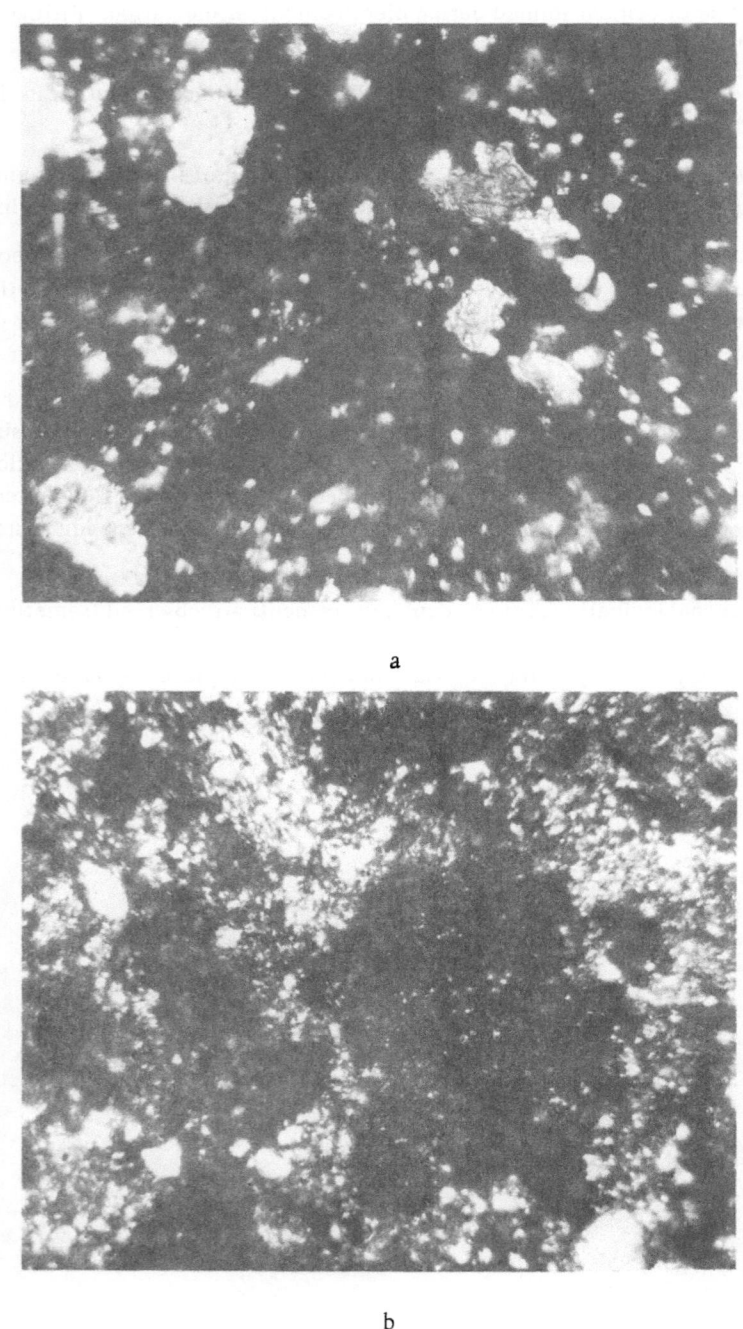

a

b

Fig. 23. Cement (× 340, crossed nicols): a) allophane (black) with idiomorphic calcite (light); b) hisingerite (black) with gibbsite (light).

THE CHANGE IN CLAY PROPERTIES DUE TO A DIRECT CURRENT

During ferruginization, which occurs widely in electrochemically indurated clays, ionic hydrates of iron are mixed with the clay, changing its color, increasing the refractive index to 1.638 or more, changing the birefringence, and altering its swelling and wetting properties. Hydrated ferrous sulfate with a refractive index < 1.540 was found in the dry residue of aqueous extracts of ferruginized areas of clay. It has a micaceous habit and a yellowish-white color, and is apparently siderotyl, the SO_4 ion for its formation being acquired from the rock.

In another type of ferruginization found in indurated rocks (in severe conditions), iron is removed neither by water nor by weak acids. These rocks undergo the greatest change during electrical treatment. For example, very marked changes were observed in carbonate-montmorillonitic clays with additions of stratal water (of the calcium chloride type), water glass and calcium chloride, combined aluminum and iron electrodes being used.

After induration, examination under the microscope frequently reveals dingy or dark areas of rock (Fig. 24), resembling "opacitization." Darkening of the clay aggregate grains is so marked that some areas become opaque between crossed nicols (Fig. 25). In reflected light the dark areas are white and porcelaneous.

A detailed study showed that this clay is impregnated by an amorphous alumina gel.

When electric current acts for about 100 hr or more, a considerable layer of clay (10-20 mm or more) is combined into a common monolithic structure. The bond is a "double" one; first a complex of mineral cements is formed (calcite, gibbsite, allophanoids, and iron hydroxides), its composition depending on the technology employed; in the second place, montmorillonitic clay is converted to ferruginous montmorillonite and nontronite, depending on the zone in which induration occurred. Formation of nontronite is due to both calcium and sodium montmorillonite, as a result of partial exchange of aluminum cations for trivalent iron.

Martitization, i.e., the formation of a hematite pseudomorph after magnetite, was observed in some experiments.

An amorphous gel, aluminum hydroxide, is formed in a medium with pH= 3.5-6.8. In experiment 139 an aluminum gel visible under the microscope and on the thermograms was formed in the anode zone for an induration period of 100-400 hr. Extensive endoeffects at 230-300°C due to the presence of free alumina, are observed on the thermal heating curves. Similar exoeffects are noted in the temperature range 900-930°C (Fig. 26), evidently owing to cubic Al_2O_3 formed during the thermal investigation. In the same experiment, after current had been passed for 1800 hr the aluminum gel at the anode disappeared, because the pH (> 3) of the medium changed. During this period the composition of the medium also changed at the cathode as a

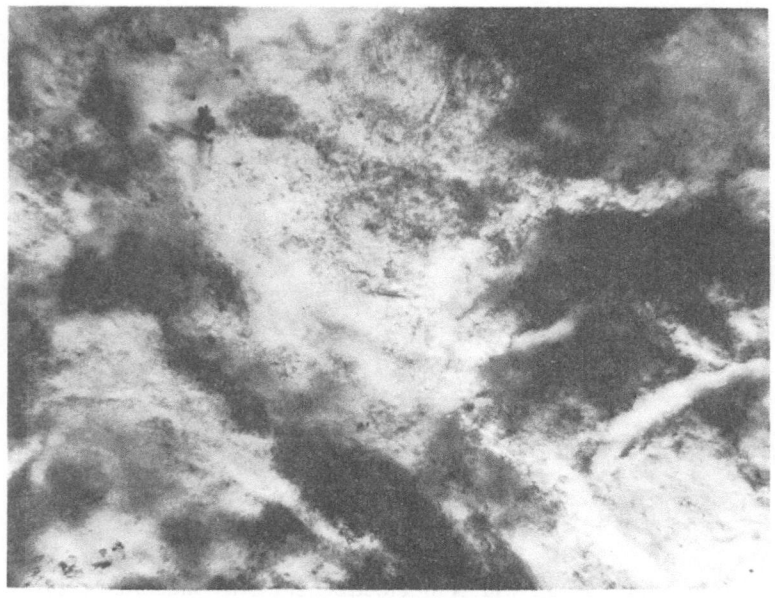

Fig. 24. Dark field — a mixture of aluminum hydroxide and clay
(× 340, parallel nicols).

result of which the pH around the latter became acid (∼5) and an aluminum gel was deposited on the gibbsite grains. It may be concluded therefore that the stable form of aluminum gel depends on the hydrogen-ion concentration.

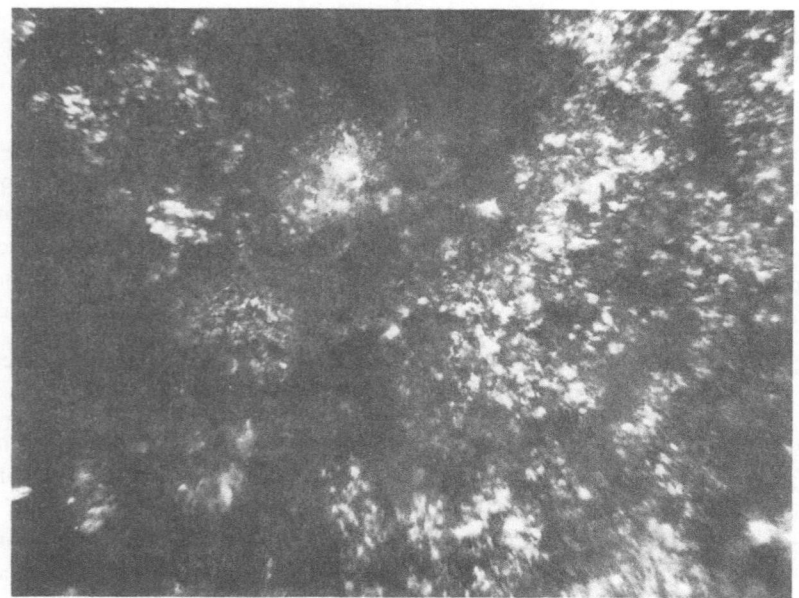

Fig. 25. Opaque areas formed by a mixture of aluminum hydroxide
and clay (× 340, crossed nicols).

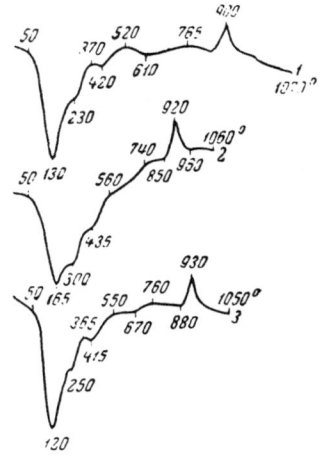

Fig. 26. Differential thermal curves of indurated bentonite (Oglanly) in the anode zone with induration for: 1) 100 hr; 2) 200 hr; 3) 400 hr.

Redeposition and recrystallization of calcite, its dissolution on the anode (acid medium), and formation on the cathode (alkaline medium) are very extensive during induration of clays.

Redeposition and recrystallization of calcite were investigated under various conditions. In some cases the experiments were performed with different monomineralic and polymineralic rocks with natural salinization, while in others a 1% calcium chloride solution was added at different electrode positions (Fe — cathode, Al — anode; Al — cathode, Fe — anode; Fe — anode and cathode; Al — anode and cathode). In the third case, different monomineralic clays were subject to artificial salinization with calcium carbonate (up to 15% wt in terms of the rock), using graphite electrodes applied to clay wetted with water. The current time was 100 hr.

In the case of montmorillonitic clay, in every experiment all the calcite was dissolved at the anode and deposited at the cathode, except for recrystallization of calcite with added calcium chloride and the following location of the electrodes: Fe — cathode, Al — anode.

In the case of salinization of montmorillonitic clay by 15% wt $CaCl_2$, not more than 3% recrystallized calcite was observed at the cathode; in the case of natural salinization of clay by calcite ($\sim$ 3%), the latter dissolved completely at the anode and was deposited on the cathode.

In the case of salinization of kaolinitic clay by 15% wt $CaCl_2$, not more than 2-3% recrystallized calcite is observed on the cathode. With natural salinization of this clay by 2% calcite, it dissolved completely at the anode and was redeposited at the cathode.

CHEMICAL-MINERALOGICAL CHANGES IN CLAYS BY CURRENT
DURING ELECTROCHEMICAL INDURATION

In the foregoing we have examined the forms in which secondary minerals are deposited, and have also given the structural characteristics of indurated rocks. We shall now examine the same processes, but from the viewpoint of the effect of induration conditions on the character of mineral formation and the chemical composition of the indurated clays. Some highly characteristic examples are given below (Tables 10, 11, and 12).

Table 10 gives the pH change in clays near the cathode and anode, and the corresponding mineral formation in monomineralic clays.

Readily exchangeable bases such as bentonite (particularly the Oglanly type) very rapidly acquire a strongly acid reaction near the anode, and an equally strongly alkaline reaction near the cathode.

Montmorillonitic clays, which also exchange bases relatively readily, sometimes acquire a high alkalinity at the cathode, but at the anode this clay is never as acid as bentonite. Kaolinite behaves relatively inertly; it is "acidified" at the anode but remains almost neutral at the cathode (only the addition of calcium salts, which migrate to the cathode, bestows an alkaline reaction on kaolinite).

The dependence of the type of new minerals formed on the clay composition and electrode character is very distinct. Experiment 9 is particularly noteworthy: in this, sodium-containing minerals, formed from material derived from the clay, were found in the cathode deposit. The alkalinity or acidity of the induration area controls the crystallization of the gels and carbonate distribution.

The change in chemical properties of a clay during induration (see Tables 11 and 12) was shown in the experiment on Jurassic and Upper Permian clay, in both the anode and cathode zones if induration was carried out without a polarity change, and in the cathode—anode and the anode—cathode zone* if there was a polarity change within one hour. The experimental conditions, electrode material, and the secondary minerals formed are also given in these Tables.

For each specimen the chemical analytical data are given in two forms: 1) the overall composition (in percentage composition at room temperature), and 2) the change in composition (by comparison with the initial clay). This column (the "change in composition" column) requires some explanation. Microscopic observations showed that silica is the component which undergoes least migration during induration; most of the clay skeleton and the quartz contained in the rock are retained; opaline material, formed by introduction

---

* In this nomenclature the first electrode mentioned is the one at which the process was commenced.

TABLE 10. Composition of the Initial and Indurated Tuimazy Clay (vol. %)

Initial clay and experiment	SiO₂	Al₂O₃	Fe₂O₃	CaO	MgO	Na₂O	K₂O	CO₂	H₂O	Additive	Electrodes	Secondary minerals
Initial clay	50.66	14.38	6.68	9.35	4.65	0.88	2.48	6.77	3.18			
Experiment 2												
Cathode	41.42	13.51	5.79	11.68	4.57	5.99	2.30	9.73	4.86	Stratal water, 1	Fe	Calcite trona (soda) (thermonatrite)
Anode	33.38	9.66	39.05	0.22	2.05	0.61	1.47	Traces	10.38	The same	Fe	Limonite (nontronite)
Change in composition												
Cathode		+1.75	+0.33	+4.04	+0.77	+5.27	+0.27	+4.19	+2.26			
Anode		+0.18	+34.65	−5.94	−1.01	+0.03	−0.15	−4.46	+8.29			
Experiment 13												
Anode-cathode	43.86	17.31	6.49	10.66	4.49	2.39	2.08	8.03	3.69	Stratal water, 1; 0.1% solution of Na₂O·SiO₂	Fe+Al	Calcite (allophane)
Cathode-anode	42.03	26.50	7.81	6.11	3.90	1.07	2.02	3.50	5.89			Gibbsite (allophane)
Change in composition												
Anode-cathode		+4.85	+0.73	+2.57	+0.46	+1.63	−0.07	+2.17				
Cathode-anode		+14.57	+2.27	−1.65	+0.04	+0.34	−0.04	−2.12				
Experiment 69												
Cathode	45.49	16.47	7.14	11.30	4.55	1.64	2.21	8.29	3.58	Dilute (1:10) stratal water +0.1% solution of Na₂O·SiO₂ +2% CaCl₂	Al+Fe	Hisingerite, calcite, allophane
Anode	47.77	16.47	16.00	4.07	4.03	1.45	2.23	2.70	4.68	The same		Hisingerite (most strongly colored allophane, hydrohematite)

									Solution	Composition
Change in composition										
Cathode	+3.56	+1.14	+2.90	+0.37	+0.85	+0.02	+2.21			
Anode	+2.91	+9.70	−4.75	−0.35	+0.62	−0.11	−3.69			
Experiment 67										
Cathode–anode	42.93	12.11	7.65	3.69	2.30	2.20	5.97	4.48	Stratal water 2+ Al+ Fe 0.1% Na$_2$O · SiO$_2$ solution	Hisingerite (calcite) (allophane)
Anode–cathode	45.06	10.89	8.87	4.14	2.50	2.18	6.72	3.95	The same	Calcite, hisingerite (allophane), hydrohematite
Change in composition										
Cathode–anode	+10.38	+6.45	−0.25	+0.25	+1.55	+0.10	+0.23			
Anode–cathode	+5.14	+4.95	+0.55	0.00	+1.72	−0.02	+0.70			
Experiment 22										
Mixed zone	44.52	8.41	7.83	4.70	1.02	2.41	12.50		Stratal water 1+ Al+ Fe 0.1% solution of Na$_2$O · SiO$_2$ + 2% CaCl$_2$	Zone 1 + 2 (hisingerite, allophane, opal substance)
Change in composition	+1.26	+2.54	−0.38	+0.61	+0.25	+0.23	+6.55			

NOTE: 1) Induration was carried out for 100 hr in all experiments except No. 22 (40 hr). 2) No. 13 was carried out without a change in polarity for the first 50 hr, and then with a polarity change.

TABLE 11. Composition of the Initial and Indurated Podol'sk Clay (vol. %)

Initial clay	SiO$_2$	Al$_2$O$_3$	Fe$_2$O$_3$	CaO	MgO	Na$_2$O	K$_2$O	CO$_2$	H$_2$O	Additive	Electrodes	Secondary minerals
Initial clay	55.55	21.08	5.96	1.54	1.81	0.35	2.59	Traces				
Experiment 111												
Cathode	41.76	18.28	10.72	8.17	1.39	3.21	1.93	4.70	0.71	1% solution CaCl$_2$+NaCl	Al+Fe	Calcite (gypsum, gibbsite)
Anode	41.76	25.78	12.48	0.75	1.70	0.61	1.74	0.10		The same	Al+Fe	Limonite
Change in composition												
Cathode		+2.40	+6.25	+7.01	+0.03	+2.95	−0.01	+4.70				
Anode		+9.90	+8.01	−0.41	+0.34	+0.35	−0.19	+0.10				
Experiment 109												
Cathode	45.39	20.32	16.12	3.30	1.45	0.37	1.80	1.67	5.49	Tap water	Al+Fe	Calcite (gypsum)
Anode	48.91	21.31	14.25	0.33	1.32	0.41	2.03	Traces	6.65	The same		Absent
Change in composition												
Cathode		+3.06	+11.26	+2.04	−0.03	+0.08	−0.32	+1.67				
Anode		+2.72	+9.01	−1.03	−0.27	+0.10	−0.28					
Experiment 112												
Cathode-anode	32.66	25.19	12.54	4.73	0.46	5.25	6.90	3.10		1% solution CaCl$_2$+NaCl	Al+Fe	Gibbsite, hisingerite, hydrohematite, calcite
Anode-cathode	33.33	20.64	13.49	4.54	0.48	6.95	0.82	5.91		The same	Al+Fe	Gibbsite (calcite), hisingerite, hydrohematite
Change in composition												
Cathode-anode		+12.39	+9.04	+3.80	−0.63	+5.04	−0.67	+3.10				
Anode-cathode		+7.92	+9.90	+3.62	−0.61	+6.74	−0.73	+5.91				
Experiment 12 a												
Cathode-anode	32.70	12.51	29.74	6.27	1.35	3.86	1.25	4.22	6.78	Stratal water 2 + 0.1% solution of Na$_2$O·SiO$_2$	Fe	Calcite, gypsum
Anode-cathode	32.47	17.12	32.51	3.02	1.57	1.85	1.31	0.78	8.29	The same	Fe	Limonite, gypsum, calcite
Change in composition												
Cathode-anode		+0.08	+26.24	+5.36	+0.29	+3.65	−0.27	+4.22				
Anode-cathode		+4.77	+29.03	+2.12	+0.51	+1.65	−0.20	+0.78				

										Metal	Mineral composition
Experiment 13a											
Cathode–anode	32.52	27.69	7.92	6.15	1.67	3.36	1.68	2.27	8.93	Stratal water 2 + 0.1% solution of $Na_2O \cdot SiO_2$	Al	1st layer: allophane (gibbsite), calcite; 2nd layer: aluminite, calcite
Anode–cathode	39.52	26.97	12.25	1.23	1.29	1.85	1.93	0.68	0.85	The same		Gibbsite (calcite, allophane)
Change in composition											
Cathode–anode		+12.66	+3.69	+5.06	+0.60	+3.11	−0.16	+2.27				
Anode–cathode		+11.94	+8.02	+0.14	+0.22	+1.60	+0.09	+0.58				
Experiment 14a											
Cathode–anode	29.96	20.37	23.29	5.87	1.39	1.66	1.45	2.22	10.15	Stratal water 2 + 0.1% solution of $Na_2O \cdot SiO_2$	Fe	1st layer: calcite, aluminite; 2nd layer: calcite
Anode–cathode	31.70	22.79	22.98	3.44	1.00	2.87	1.46	2.99	9.25	The same	Fe	1st layer: limonite, calcite, allophane; 2nd layer: no secondary minerals
Change in composition											
Cathode–anode		+8.93	+20.08	+5.04	−0.41	+1.47	+0.06	+2.22				
Anode–cathode		+10.74	+19.59	+2.56	−0.03	+2.67	+0.01	+2.99				

NOTE: In all cases the induration time was 100 hr, with a change in polarity in all experiments except 109 and 111.

TABLE 12. Technological Conditions of Treatment of Monomineralic Rocks by Direct Electric Current

Bentonite (Oglanly)

Expt. No.	Electrodes	Induration time, hr	Solutions	Current direction	pH Cathode	pH Anode	Current density, ma/cm²	Voltage, V	Secondary minerals Cathode	Secondary minerals Anode
140	Graphite in anode and cathode	100	H_2O	Without a change in polarity	>12; 12	1.1–1.4	61	65.8	Calcite*	Graphite
158		15		The same			32	26.2	Aluminum gel	Absent
167		15		With change in polarity	5–11.5–13	1.1	63	39.0	Calcite, † limo-nitc†	Al gel
150	Al-Anode Fe-Cathode	15	1% $CaCl_2$	Without a change in polarity	12, >12	1.2	22	48.3	Al gel, calcite†	Allophane,† calcite†
151		100		The same	9.7 and >12	11.7–11.5	25	47.5	Allophanoid, hisingerite,† calcite†	Allophanoid,† Al gel†
141	Graphite	100	15% $CaCO_3$ (solid)	Without a change in polarity	>13.0	−1.8 – 2.0	66	64.8	Calcite	Graphite
139	Al-Cathode Al-Anode	100	1% $AlCl_3$	The same	12, >12	1.6–2.3	46	57.43	Gibbsite, calcite*	Al and Cu gel
146			1% NaCl solution	"	11.7	2.5	24	37.8	Gibbsite, allophanoid	Allophanoid, calcite,† hisingerite,‡
161		15	Addition of 1% $AlCl_3$	"	9.0	3.4	28	62.2	Metallic gibbsite calcite†	Allophanoid, gibbsite,† Al gel*
169	Al-Cathode Al-Anode	15	H_2O	Without a change in polarity			44	45.1	Calcite† Al gel	Limonite† Fe adsorbed by clay, gibbsite† in ferruginized area
178		15	1% NaOH 1% $AlCl_3$	The same	10.75	4–3.8	44	46.4	Middle gibbsite, allophanoid,† hisingerite,* calcite*	

No.	Electrodes		Solution	Polarity	pH				Secondary minerals	
138	Fe-Cathode Fe-Anode	100	1% $FeCl_3$	Without a change in polarity	12, >12	1.6–2.3	39	61.6	Magnetite,* calcite,* hydrohematite*	Ferruginized clay, ferruginized calcite,† hydrohematite†
168		15	1% $CaCl_2$	The same			47	36.6	Magnetite, ferruginized clay, limonite,* hydrohematite*	Magnetite, hydrohematite,* limonite,* ferruginized clay
141	Graphite	100	1% $CaCl_2$	Without a change in polarity	7.5–11.3	0.8	74	60.8	Calcite, ferruginized clay	Graphite
173	Al-Anode Fe-Cathode	15	H_2O	The same	11–11.7	3.8	25	45.2	Al gel	Al gel
143	Graphite	100	1% $FeCl_3$	"	>12, >13	2.3–2.5	50	58.3	Magnetite,* hydrohematite,* calcite*	Ferruginized clay
144		100	1% $AlCl_3$	"	10.8–11.2	0.75–1.0	38	78.9	Allophanoid, gibbsite,† calcite*	Allophanoid,* Al gel
145	Graphite	100	1% $Al(OH)_3$ suspension	Without a change in polarity	> 12.5 middle 9.75	7.7	76	58.6	$Al(OH)_3$, Al gel, calcite*	$Al(OH)_3$, Fe gel
147	"	100	1% $Al_2(SO_4)_3$ ·$18H_2O$	The same	2.0–1.6	0.1–0.3	41	37.8	Hydrohematite, calcite,* allophanoid,† hisingerite‡	Clay unchanged
148	"	100	1% NaOH	"	11.2–11.5–11.7	1.75–3.1	30	37.8	Na carbonates, Al gel, allophanoid‡	Iron adsorbed by clay
149	"	100	3% $AlCl_3$	"	1.1	0.6	47	37.8	Allophanoid,* gibbsite,‡ iron group†	No secondary minerals

Monothermite (Chasov-Yar)

No.	Electrodes		Solution	Polarity	pH				Secondary minerals	
154	Graphite	100	H_2O	Without a change in polarity	10.2–10.5	1.75–2.0	10	37.8	Calcite,† opal substance†	Graphite, Al gel
9	Al-Anode Fe-Cathode	15	1% $CaCl_2$	The same	11.59	4.03	23	30.0	No secondary minerals found	NaCl $CaCl_2$

TABLE 12. Technological Conditions of Treatment of Monomineralic Rocks by Direct Electric Current (continuation)

Expt. No.	Electrodes	Induration time, hr	Solutions	Current direction	pH Cathode	pH Anode	Current density, ma/cm²	Voltage, V	Secondary minerals Cathode	Secondary minerals Anode
4	Al-Anode Fe-Cathode	100	1% CaCl₂	Without a change in polarity		7.18	17	30.6	Calcite, sodium carbonate, Fe of adsorbed clays	Calcite and chloride salts
8	The same	30	"	The same	11.50	4.19	30	30.0	γ-Gibbsite, calcite*	No secondary minerals
12	" "	5	"	"	8.9	4.2	45	30.0	Fragments of iron electrode	No secondary minerals
2	One electrode Al, the other Fe	100	—	With a change in polarity	—	—	35	30.3	Al-zone, gibbsite, γ-gibbsite, allophanoid	Fe-zone, magnetite
3	The same	30	1% CaCl₂	With a change in polarity	—	—	30	30.5	γ-Gibbsite,†	Hydrohematite, magnetite
6	" "	15	"	The same	—	—	47	30.0	Gibbsite, calcite*	Fe-impregnated clay,† limonite,† hydrohematite†
182	Graphite	100	15% CaCO₃ (solid)	Without a change in polarity	12.3	0.5	17	61.5	Granular calcite, Al gel	Granular calcite, Al gel, recrystallized calcite*
5	Al-Cathode Fe-Anode	100	1% CaCl₂	The same	11.2	3.9	18	30.5	Gibbsite, magnetite,* calcite*	Fe-impregnated clay, limonite,* hydrohematite*
7	The same	30	"	"	11.8	2.9	38	30.0	Gibbsite, γ-gibbsite, calcite*	Fe-impregnated clay
10	" "	15	"	"	10.3	2.7	15	35.0	Calcite, γ-gibbsite	Fe-impregnated clay
13	" "	5	"	"	10.0	3.3	35	30.0	γ-Gibbsite	Fe-impregnated clay

Kaolinite (washed)

No.	Electrode	Medium	Amount	Polarity	pH	pH		%		
155	Graphite	H_2O	100	Without a change in polarity	7.9-9.2; (11.3)	1.0	15	37.2	Calcite*	Fe adsorbed by clay
152	Al-Anode Fe-Cathode	1% $CaCl_2$	15	The same	7.6-8.7	1.1	21	48.3	Calcite* (crust)	Al gel, allophanoid,* hisingerite†
153	The same	"	100	"	10.1-10.2	10.5-1.2	10	47.5	Calcite, hisingerite,* allophanoid†	Calcite,* hisingerite,† allophanoid†
157	Combination of Al+Fe+Al	1% $CaCl_2$	15	"			14	52.5	Calcite*	Hydrohematite,* limonite,* ferruginized clay

Zai-Karatai bentonite

No.	Electrode	Medium	Amount	Polarity	pH	pH		%		
170	Graphite	H_2O	15	Without a change in polarity	9.1	4.1-4.5	12	51.0	—	—
171	"	1% $CaCl_2$	15	The same	>9	3.0	34	41.4	Calcite	—

*Considerable amount.
†Few not more than 1%.
‡Very few of individual grains.

NOTE: Graphite obtained by disintegration of graphite electrodes.

of water glass, is very seldom deposited in the secondary minerals (see experiment 154, Table 12). From this relative constancy of the silica component we assumed that all (or nearly all) the silica of the initial rock is retained in indurated rock, differences in the latter's composition being due to migration of the more mobile oxides. Such migration obviously occurred because the volume of the indurated soil was changed.

The assumption that the silica content is relatively constant enabled us to calculate the movement of the other oxides during each experiment. The basis for this calculation was the overall compositions of the initial and indurated clays; starting from the composition of the initial rock, we can let

$$\frac{SiO_{2\,init}}{SiO_{2\,ind}} = \frac{\text{oxide of the initial clay}}{X \cdot \text{the same oxide of indurated clay}} .$$

Thus we can easily obtain figures showing the content of a given oxide which the initial clay would have had if all the silica in the initial clay were in the combined form. Subtracting these calculated figures from the indurated clay composition, we obtain the difference, indicating either introduction of oxide into the indurated zone (shown by a figure with a plus sign in the "change in composition" column) or removal of oxide (minus sign).

Despite the arbitrariness of such calculations, the results indicate that alumina migrates extensively and becomes concentrated in the cathode zone — a fact not allowed for in natural processes.

Calcium salts very often migrate, being removed from the anode zone and deposited in the cathode zone.

In induration with a change in polarity, the predominance of deposition processes over removal processes is very interesting. In this case the quantitative ratios are fairly complex and specific for each experiment. The mineral changes in clay composition lead to the appearance of secondary minerals, mainly determined by the chemical properties of the zone; the material composition is averaged, without obtaining markedly separate zones (as may be seen in experiments without a change in polarity). However, the amount of oxides increases in both zones as a result of dissolution of the electrodes and additives.

CONCLUSIONS

1. The authors demonstrate the processes of artificial mineral formation based on chemical reactions (solution, oxidation, reduction, and hydrolysis), together with accompanying physicochemical processes of pore settling from a saturated solution (coagulation, absorption capacity, and exchange reactions, migration, and adsorption).

2. Formation of new minerals is the result of a combination of all the processes taking place during the electrochemical induration of rocks.

3. The minerals formed are artificial cements for rocks indurated by the electrochemical method.

4. It has been established that the secondary minerals, by crystallizing in the interstices between clay aggregates, cement them, forming new rock structures.

5. The predominant forms of artificial cement are basal cement and filling cement. The material composition is monomineralic crystalline or amorphous. The cements consist of gibbsite, calcite, limonite, hydrohematite, hydrogoethite-lepidocrocite, and the allophanoid group (allophane and hisingerite).

6. Induration of rock is due to the formation of cements (crystallites), which convert the rock to a monolithic structure.

7. The strength indices are closely related to the chemical and material composition of the indurated rocks. Furthermore, the course of the induration process can be controlled.

8. Direct current is only necessary for initiating crystallization. A number of experiments have confirmed that, after the current is switched off, crystallizaion of secondary minerals continues and the strength of the entire clay structure increases.

9. Secondary mineral formation processes are perfectly controllable; they may be used to investigate similar processes in natural conditions, and for strengthening weak rocks during borehole sinking and various engineering and hydrotechnical constructions.

LITERATURE CITED

1. Casagrande, L.,Die Bautechnik, No. 1 (1937).
2. Endell, K., Die Bautechnik, No. 48 (1935).
3. Erlenbach, L., Die Bautechnik, No. 19 (1936).
4. Rzhanitsyn, B. A., "Electrochemical induration of soils " in Coll.: Hydrogeology and Engineering Geology, No. 5 (1940).
5. Rzhanitsyn, B. A., "Electrochemical induration of clay soils," Conference on induration of soils and rocks, Dokl. Akad. Nauk SSSR (1941).
6. Rel'tov, B. F. and Novikov, A. V.,"Electrochemical induration of clay soils," Communications I and II, Izv. B. E. Vedeneev Hydrotechnical Research Institute,Vols. 30 and 31. (Moscow —Leningrad, 1941, 1946).
7. Tolstopyatov, B. V., "Electrochemical induration of clay soils," Pochvovedenie, No. 8 (1940).
8. Titkov, N. I., Korzhuev, A. S., Smolyaninov, V. G., Nikishin, V. A., and Neretina, A. Ya.,Electrochemical Induration of Weak Rocks (Moscow, Gostoptekhizdat, 1959). English translation: Consultants Bureau (New York, 1961).
9. Titkov, N. I., Petrov, V. P., and Neretina, A. Ya., "Minerals and structures formed in wet clay rocks by direct electric current," Report to the Conference of the International Committee on the Investigation of Clays,Izd. Akad. Nauk SSSR (Moscow, 1960).
10. Solntsev, D. I. and Borkov, V. S., "Electrochemical induration of clay soils for combating subsidence of railroad tracks," Dokl. Akad. Nauk SSSR (1941).
11. Nikolaevskii, F. A., "Data on the mineralogy of the environs of Moscow," Izd. Rossiisk. Akad. Nauk Vol. 3, No. 11 (Moscow, 1912).
12. Feodot'ev, K. M. and Starostina, V. G., "Natural and synthetic hydrated ferroaluminum silicates ", Problems of Petrography and Mineralogy, Vol. II.,Izd. Akad. Nauk SSSR (Moscow, 1955).
13. Levando, E. P., Chemical-Mineralogical Classification of Gibbsite-Boehmite Bauxites of the Tikhvin Type (Moscow, Gosgeolizdat, 1956).
14. Kitaigorodskii, A. I., X-ray Structural Analysis (1952).
15. Rooksby, H. P., Oxides and Hydroxides of Aluminum and Iron X-ray Identification and Crystal Structure of Clay Minerals (London, 1951).
16. Belyankin, D. S., Butuzov, V. P., and Feodot'ev, K. M., "On a distinctive feature of the heating curve of hydroargillite," Problems of Soviet Soil Science, Coll. 15 (1949).
17. Belyankin, D. S. and Ivanov, V. I., "Conversions of kaolin during heating," in the book: The 50th Anniversary of the Scientific and Pedagogic Activity of Academician Vernadskii, Izd. Akad. Nauk SSSR (Moscow —Leningrad, 1936).

18. Belyankin, D. S. and Feodot'ev, K. M., "The heating curve of kaolin in the light of present knowledge," Dokl. Akad. Nauk SSSR, Vol. 65, No. 3 (1949).

19. Belyankin, D. S. and Feodot'ev, K. M., "Kaolin and allophanoids; thermo-optical data and shrinkage phenomena during heating," Zap. Vses. mineralog. obshch., No. 2 (1951).

20. Rinne, Z., "Röntgenograhische Untersuchungen an einigen feinzerteilten Mineralien, Kunstrpodukten und diehnen Gesteinen," Z. Krist., Vol. 60 (1924).

21. Berestneva, Z. Ya., Koretskaya, T. A., and Kargin, V. A., "The formation mechanism of colloidal particles of aluminum hydroxide," Kolloid. zhur., Vol. 5, No. 8 (1951).

22. Sedletskii, I. D., "Classification of minerals of the erosion crust," Sov. geol., No. 3 (1941).

23. Nikolaevskii, F. A., "Allophanoids from the environs of Moscow," Izd. Rossiisk. Akad. Nauk, ser. 6(1912).

24. Chukhrov, F. V., Colloids in the Earth's Crust, Izd. Akad. Nauk SSSR (Moscow, 1955).

25. Rozhkova, E. V. and Lyamina, A. N., "Diospore from bauxite deposits of the USSR," Tr. Vses. nar. inst. mineral'n. syr'ya, No. 2 (1949).

26. Mokievskii, V. A., Stulov, N. N., and Tsigel'man, I. S., "Mineral formation in a natural electric field," Zap. Vses. mineralog. obshch. Part 35, No. 1 (1956).

27. Lepin', L. K., Vaivade, A. Ya., and Oshis, Z. F., "Relation between the corrosion rate of iron and solution pH and metal passivation in an alkaline medium," Zh. fiz. khim., Vol. 29, No. 2 (1955).

28. Smirnov, S. S., The Oxidation Zone of Sulfide Deposits, Izd. ONTI NKTP (Moscow—Leningrad, 1936).

29. Titkov, N. I., Petrov, V. P., and Neretina, A. Ya., "Mineralogical-petrographic methods of investigating electrochemically indurated rocks." in Coll.: Experimental Investigations at Working Sites of Deep Oil and Gas Deposits, Izd. Akad. Nauk SSSR (Moscow, 1964).